Make:

ReMaking History

Makers of the Modern World

Volume **3**

Make:

ReMaking History
Makers of the Modern World

William Gurstelle

MAKER MEDIA
SAN FRANCISCO, CA

ReMaking History, Volume 3
Makers of the Modern World
By William Gurstelle

Printed in Canada.

Published by Maker Media, Inc.,
1160 Battery Street East, Suite 125
San Francisco, California 94111

Maker Media books may be purchased for educational, business, or sales promotional use. Online editions are also available for most titles (*safaribooksonline.com*). For more information, contact our corporate/institutional sales department: 800-998-9938 or *corporate@oreilly.com*.

Publisher: Roger Stewart
Editor: Roger Stewart
Copy Editor: Rebecca Rider, Happenstance Type-O-Rama
Proofreader: Elizabeth Welch, Happenstance Type-O-Rama
Interior Designer and Compositor: Maureen Forys, Happenstance Type-O-Rama
Illustration: Richard Sheppard, Happenstance Type-O-Rama
Cover Designer: Maureen Forys, Happenstance Type-O-Rama
Indexer: Valerie Perry, Happenstance Type-O-Rama

February 2017: First Edition

Revision History for the First Edition
2017-02-01: First Release

See *oreilly.com/catalog/errata.csp?isbn=9781680450729* for release details.

978-1-68045-072-9

Safari® Books Online

Safari Books Online is an on-demand digital library that delivers expert content in both book and video form from the world's leading authors in technology and business.

Technology professionals, software developers, web designers, and business and creative professionals use Safari Books Online as their primary resource for research, problem-solving, learning, and certification training.

Safari Books Online offers a range of plans and pricing for enterprise, government, education, and individuals. Members have access to thousands of books, training videos, and pre-publication manuscripts in one fully searchable database from publishers like O'Reilly Media, Prentice Hall Professional, Addison-Wesley Professional, Microsoft Press, Sams, Que, Peachpit Press, Focal Press, Cisco Press, John Wiley & Sons, Syngress, Morgan Kaufmann, IBM Redbooks, Packt, Adobe Press, FT Press, Apress, Manning, New Riders, McGraw-Hill, Jones & Bartlett, Course Technology, and hundreds more. For more information about Safari Books Online, please visit us online.

How to Contact Us

Please address comments and questions concerning this book to the publisher:

Maker Media, Inc.
1160 Battery Street East, Suite 125
San Francisco, CA 94111
877-306-6253 (in the United States or Canada)
707-639-1355 (international or local)

Maker Media, Inc. unites, inspires, informs, and entertains a growing community of resourceful people who undertake amazing projects in their backyards, basements, and garages. Maker Media, Inc. celebrates your right to tweak, hack, and bend any technology to your will. The Maker Media, Inc. audience continues to be a growing culture and community that believes in bettering ourselves, our environment, our educational system—our entire world. This is much more than an audience; it's a worldwide movement that Maker Media, Inc. is leading and we call it the Maker Movement.

For more information about Maker Media, Inc. visit us online:

- *Make:* magazine *makezine.com/magazine*
- Maker Faire *makerfaire.com*
- Makezine.com *makezine.com*
- Maker Shed *makershed.com*
- To comment or ask technical questions about this book, send email to *bookquestions@oreilly.com*.

Please visit *www.ReMakingHistory.info* for information including corrections, updates, and further reading recommendations.

Acknowledgments

Thanks to the crack editing and production team at Happenstance Type-O-Rama for their hard work.

Sincere appreciation to the staff at Maker Media for all they've done.

Thanks to my wife Karen for her invaluable help and support.

About the Author

William Gurstelle has been writing for *Make:* magazine pretty much since the beginning. Besides 40 or so "ReMaking History" columns, his work there has included do-it-yourself pieces on a gravity-powered catapult, a taffy-pulling machine, a Taser-powered potato cannon, and an ornithopter. He's also a bestselling author, a registered engineer, and a popular speaker on the world of science and technology.

Contents

The Golden Age of Invention

Try to imagine what life was like 150 years ago, before many of the great modern inventions that shape our world came on the scene. Imagine a time when the only music you could hear was played live—a time before airplanes allowed us to span continents in hours, not months or years. Imagine a time when all news was local because any event that took place more than a couple of days' ride away by horseback was too far away to care about.

For better or worse (and in almost everyone's opinion, it's certainly for the better), technology has mightily changed the way we wander through our day-to-day lives, from the moment we're born to the time we leave this world.

This book explores the incredibly rich inventive time period that roughly spans the beginning of the 19th century to the middle of the 20th. It can be argued that this window of time saw the most rapid transformation of society due to technology. If a person living in say, 1600, were time transported to 1750, he or she would notice a great many astounding changes in the world, without a doubt. But that person, I argue, would be far more able to understand and would much more quickly adapt to his or her surroundings than a person who teleported from 1800 to 1950.

That's because of the breathtaking pace of change that was a hallmark of this period that I term the *Golden Age of Invention*. Automobiles, airplanes, refrigeration, electronic communications, movies, and radio are just a few of these Golden Age of Invention technologies.

Undoubtedly, the people of the time were aware of the breakneck pace at which technology was changing their world, and they had to try mightily to understand what they saw going on around them. Imagine moving from horses to cars, from kerosene lanterns to electricity, and from, well, nothing to radio in just a few decades.

In some quarters, people genuinely feared that invention and technology were changing the world too quickly, and in ways that were not good for many. But for a great many others, the Golden Age of Invention was an exhilarating, even thrilling time to be alive.

Every year brought something radical and new, and these inventions were greeted with terrific interest. The scientists and inventors of that period were rock-star famous. Many were A-list celebs known to the average person on the street the way that the best NFL quarterbacks and movie stars are today.

Although a few of today's inventors (say, Steve Wozniak or Tim Berners-Lee) are well known, they don't garner anywhere near the fame that people like Golden Age inventors Samuel Morse, George Washington Carver, or Alexander Graham Bell enjoyed during their lifetimes. These inventors' names were known by young and old, and people followed their latest inventions closely.

World's Fairs and Expos

Today, to find out about the latest mobile telephone or smart automobile technology, you need do nothing more than type in a few well-formed phrases in the search bar of your Internet browser, download a podcast, or tune in a technology program on the television. But during the Golden Age, there was no easy way to satisfy your curiosity about the latest and greatest gadgets. Often, you had to go and look at the invention in person. Since products were made all over the world, it was hard to stay abreast of the times. But there was one way to learn a great deal in a short time. You could attend a world's fair.

Although they still exist, albeit in forms that are but shadows of their former glory, world's fairs or world expositions were once the largest, most well attended, and most discussed public events on earth. The first real world-class event was staged in London in 1851, and there have been a couple dozen really big fairs since then, depending on your definition of "really big." But the heyday of world's fairs pretty much coincided with the Golden Age of Invention. From the about the middle of the 19th century to the middle of the 20th century, people thronged to world's fairs and expos, eager to see and experience the latest technologies and inventions.

Fairs were (and still are, in parts of Asia) unbelievably huge and expensive undertakings. They cost millions of dollars to put on, and in many cases, a large proportion of the population of the host country and surrounding countries showed up. For example, the attendance figures for the Exposition Universelle, an international fair in Paris in 1900, were on the order of 50 million visitors! At the time, the country of France's entire population was about 38 million people. That's a lot of visitors!

Why were fairs such a big deal? Because, during the Golden Age of Invention, there was no better way to show many people at the same time the promise that technology held for their future. In the days before television and the Internet, the way to do that was in a carefully planned city of the future designed just for that purpose.

World's fairs were more than simply giant exhibitions of manufactured goods and architecture; they were also venues for friendly competition between countries. Such contests staked national pride upon how well the technological prowess of one country matched up with another. Fair organizers selected groups of expert evaluators to judge exhibits and award prizes to the very best. Taking home a blue ribbon was a source of great corporate and national pride.

For the average technophile, a fair was a glorious opportunity to walk among futuristic and fantastic architecture while also touring the latest and greatest ideas the world had to offer, all without needing to buy an airplane or steamship ticket. In just a few days, one could encounter many of the newest inventions of the time. World's fairs promised the future, and more often than not, delivered on their promise.

The Exposition Universelle de 1889, also in Paris, debuted a host of new industrial inventions that would soon light and power the world, such as motors, transformers, dynamos, and conveyors, as well as consumer innovations, such as escalators, sound films, and an Edison phonograph. The phonograph was an especially big hit as it wowed the crowds by alternately playing the *Star Spangled Banner* and *La Marseillaise* over and over again. Remember, up to that point, if you heard music, there was a group of real-live persons playing it. To see and hear a machine doing it must have been almost unbelievable.

Four years later, the 1893 Chicago World's Columbian Exposition unveiled the dishwasher, spray paint, the zipper, the Ferris Wheel, Cream of Wheat, Shredded Wheat, and Cracker Jack. Some historians estimate that one out of every four Americans then alive visited the Chicago World's Fair. That's beyond amazing, considering the cost and difficulty of interstate travel at that time, and further, that the event lasted only six months.

The 1939 New York World's Fair in Flushing Meadow featured robot Elektro (a seven-foot-tall robot that could walk, talk, and smoke cigarettes,) nylon fabric, fax machines, and the very first science fiction convention. Most significantly, the age of television began with a telecast of President Franklin D. Roosevelt opening the fair with a welcome broadcast on April 30, 1939. At the time, there were only 200 television sets in all of New York City.

How This Book Is Organized

As you turn these pages, you might imagine yourself strolling through the pavilions of a bygone world's fair. Each chapter spotlights a great thinker or inventor from the past and a description of what that person did to make it into the history books. The Golden Age of Invention, from roughly 1800 to 1950, was a particularly rich period, both in terms of interesting inventions and the interesting personalities of the people who thought them up (see Figure I.1).

First, we'll take a brief look at the person and then we'll examine their claim to fame—that is, the wonderful or important thing they invented or discovered, the nature of the science behind it, and how it made the world a better place. Finally, and this is the best part, we'll build a simplified version of the invention so we can really understand it and see for ourselves how and why it works.

By the way, this book is part of a series. Besides looking at the works of these Golden Age inventors, the other books in this series examine the contributions of very early inventors from ancient civilizations as well as more recent famous inventors from the days of the Industrial Revolution like Benjamin Franklin, Charles Goodyear, and Humphry Davy.

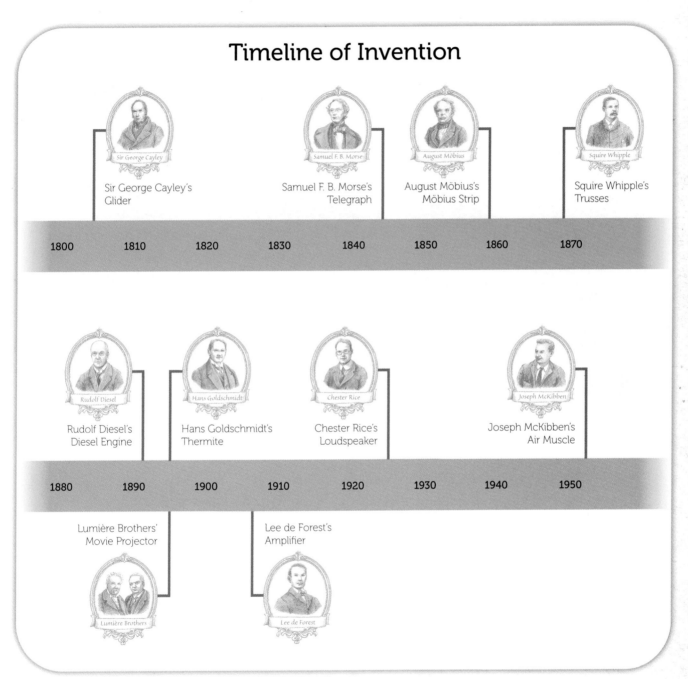

Figure I.1: Timeline of Invention, the Industrial Revolution

First Things First: Working Safely

The projects described in the following pages have been designed so you can make and use them as safely as possible. However, as you try them out, there is still a possibility that something unexpected may occur. Some of the projects involve the use of things that are hot, heavy, or sharp, and you need to be careful when you work with these items. It is important that you understand that neither the author, the publisher, nor the bookseller can or will guarantee your safety. When you try the projects described here, you do so at your own risk.

These are your general safety rules. You will also find that some chapters provide specific safety instructions for the particular project or experiment in that chapter.

1. Read the entire project description carefully before you begin the experiment. Make sure you understand what the experiment is about and what it is that you are trying to accomplish. If something is unclear, reread the directions until you fully comprehend them.

2. Wear protective eyewear, gloves, and so on, when indicated in the directions.

3. The instructions and information are provided here for your use without any guarantee of safety. Each project has been extensively tested in a variety of conditions. But variations, mistakes, and unforeseen circumstances can and do occur; therefore, all projects and experiments are performed at your own risk. If you don't agree with this, then put this book down; it is not for you.

4. Finally, believe me when I tell you that it's no fun getting hurt. I want you to stay in one piece. And the very best way to do that is to use your own common sense. If something doesn't seem right, stop and review what's happening. (That doesn't just pertain to what's in this book; that's my advice to you in general.) You must take responsibility for your personal safety and the safety of others around you.

Getting Started

Inventions can come about in two ways: by discovery or by recombination. Inventions that are completely new are rare. Many more are the result of ingenious and clever recombination of stuff that's already there.

When an English baronet from rural Yorkshire named George Cayley discovered the basis for modern airplane flight, he was completely on his own. In a flash of rare brilliance, he figured out that there were four aerodynamic forces involved in flying—thrust, lift, weight, and drag—and making a thing fly was merely an exercise in juggling those forces. Every airplane design from the Wright brothers on has been based on this discovery.

But completely new and pioneering ideas such as Cayley's are uncommon. More common by far are inventions of recombination; they comprise the bulk of the projects in this book. Lee de Forest's vacuum tube, for example, was not so much the product of a new and unique discovery, as it was the insightful combination of preexisting components—electrical potential, a resistance heater, and an evacuated glass tube. None of these things was novel in the 1900s, but de Forest's ingenuity made the combination of these items a world-changing invention.

As you move through the book, you'll encounter quite a few inventors whose names you likely already know. Many people are familiar with famous scientists and inventors such as de Forest, Rudolf Diesel, and Samuel Morse. But there are at least a few, I'll wager, who have names that don't ring a bell. Chester Rice, Joseph McKibben, and Squire Whipple are hardly household names, but their inventions were and still are important and interesting to explore and replicate.

Part I

Blue Ribbon Projects

If you look, you'll see numerous echoes of the Golden Age of Invention's great world's fairs in the first several projects in this book. Consider Squire Whipple's iron truss bridges described in Chapter 3. To a civil engineer, the simple trusses in the original 1840-era Whipple bridges are the close cousins of the artistic steel lattices used in the Eiffel Tower, which was built as the entrance to the 1889 Paris Expo (Exposition Universelle de 1889) grounds.

At the 1939 New York World's Fair, RCA showed the world its latest and greatest technology by inviting the public into its huge exhibition hall. The hall, from a bird's-eye view, looked like a giant vacuum tube—the world-changing invention of Lee de Forest, a scientist we'll meet in Chapter 4.

Our first project is based on perhaps the biggest idea to come out of the 1889 Paris Expo. The diesel engine made quite a splash when it was first shown to the public there, and it has remained important ever since. When you deconstruct the workings of the modern diesel engine, what will you find at its core? A simple device called a *fire piston*. In fact, our fire piston project is a smaller, simpler version of Rudolf Diesel's internal combustion engine.

Over the centuries, people have devised many ways to kindle fire when and where they need it. Fire drills, burning lenses, and flint and steel are all dependable, if time-consuming, methods of fire starting. One of the quickest and most elegant methods is the fire piston or fire syringe.

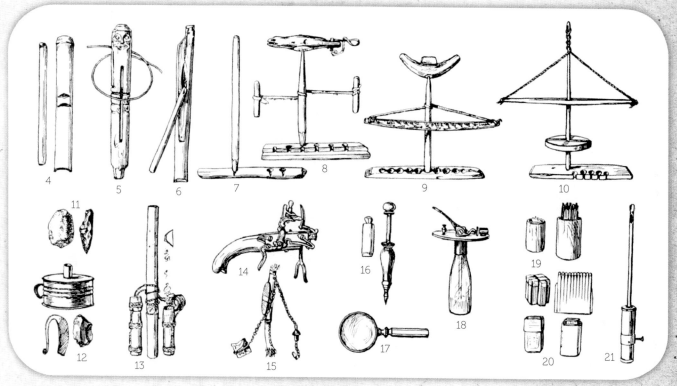

4. Fire saw. Strip of bamboo drawn across a section of bamboo. Dyaks of Borneo, Malays.
5. Fire thong. Rattan thong drawn over a grooved piece of wood. Dyaks of Borneo.
6. Fire plow. Blunt stick worked along a groove in a lower stick. Polynesians.
7. Fire drill. Slender rod twirled between the hands upon a lower stick having a cavity with slot. Indians of the United States and widely diffused in the world.
8. Fire drill. Rod held in a socket and gyrated by means of a cord. The lower piece of wood has a cavity with slot, opening upon a shelf. Eskimos of Alaska.
9. Fire drill. Rod held in a socket and gyrated with a bow and cord. Lower piece with cavities on a central groove. Eskimos of Alaska.
10. Fire drill. Pump drill used specially for sacred fire. Iroquois Indians, Canada.
11. Strike-a-light. Flint and iron pyrites struck together as the ordinary flint and steel. Eskimos of Alaska.
12. Strike-a-light. Flint and steel and box for holding flint, steel, and tinder. Sulfur-tipped splint ignited from the tinder. England.
13. Strike-a-light. Bamboo tube and striker of pottery used as flint and steel. Two boxes for tinder. Malay.
14. Tinder pistol. Gunlock adapted for throwing sparks into tinder. England.
15. Strike-a-light. Combination of flint, steel, tinder, and extinguisher, for carrying in the pocket. Spain.
16. Fire syringe. Cylinder with closely fitting piston bearing tinder. Driving the piston down smartly kindles the tinder. Siamese and Malays.
17. Lens. Used for producing fire by focusing sunlight upon tinder. Ancient Greeks.
18. Hydrogen lamp. Hydrogen gas is made, to play upon spongy platinum, causing it to glow. Germany.
19. Match light box. Bottle of sulphuric acid, into which splints tipped with chlorate of potash and sugar were dipped. Vienna.
20. Matches. Various kinds of phosphorus matches.
21. Electric gas lighter. Cylinder containing a small dynamo run by pressure of the finger, producing sparks between the points at the upper end of the tube.

Figure 1.1: How to assemble a fire piston

Rudolf Diesel

Rudolf Diesel and the Fire Piston

The project described in this chapter, a sort of primitive fire lighter, is a wonderful example of physics in action. By quickly compressing air inside a cylinder, we can raise its temperature enough to start a small bit of fuel stuffed inside it on fire.

It sounds impossible, but with a few easy-to-find parts, we can make a device that produces fire from nothing but air!

A New Kind of Automobile Engine

Rudolf Diesel was one of the youngest students ever to enroll in the mechanical engineering program at the München Polytechnikum. During his studies at this prestigious German university, the precocious student met Professor Carl von Linde, who was famous for developing refrigeration and gas separation technologies. Rudolf assisted Carl in his lab and became one of his most trusted assistants.

In 1871, von Linde returned to the university after a lecture tour that took him to Malaysia. Because this was the late 19th century, the voyage had taken him months. Von Linde had seen and learned much during his excursion to Southeast Asia, and as a faculty member of the prestigious Technical University of Munich, he was obligated to deliver a presentation on the results and findings of his trip to students and faculty. Diesel attended the lecture and was quite eager to find out what his mentor had learned during his time away.

The talk was a lengthy one and during his lecture, the fatigued Herr Doktor felt the need for a nicotine hit. He paused and withdrew from his pocket a small wooden cylinder and plunger, which he called *eine Feuerpumpe*. The small device was a present from the people he had met on Penang Island in the Straits of Malacca. The indigenous people of the region used it to start fires. A person experienced in using *die Feuerpumpe*, or fire piston, could reliably provide hot glowing embers anytime they were needed, even under the humid conditions of the rain forest.

At the lectern, Linde slapped the plunger down and the tinder inside ignited. He plucked out a glowing ember and lit his cigarette with it. It was a neat gesture; to the audience, it looked like he had produced fire from nothing at all—they saw no match, no flint. The fire had magically appeared from the bottom of what they saw as an empty, hollowed-out tube.

The demonstration lit not just a cigarette, but an idea in Rudolf Diesel's head. Diesel had been experimenting with the recently invented internal combustion engine and was growing frustrated with the spark-ignition cycle engine's inherent low efficiency. When von Linde lit that cigarette, a question jumped into Diesel's head: "Could the same thermodynamic process that ignited the tinder in the bottom of the fire piston also ignite fuel in an internal combustion engine?" If so, perhaps here was a way to significantly improve the efficiency of this type of engine. And as history proved, it was indeed.

Making a Fire Piston

You can start a fire by using chemical energy (a match), by using friction (rubbing sticks together), or by a phenomenon called *pyrophoricity* (which is the basis for flint and steel fire starters). But one of the most satisfying ways to start a fire is by quickly compressing air using a fire piston.

There are a lot of different names for the fire piston. It's also called a fire syringe or a slam rod fire maker. But no matter what you call it, it is certainly a unique way of lighting your campfire! The following steps will explain how to make your own fire piston, the antecedent of the diesel engine. Take a look at Figure 1.1 to see what you're shooting for.

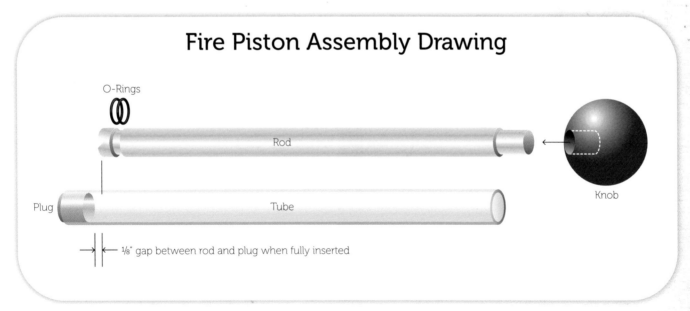

Figure 1.1: How to assemble a fire piston

Materials

(2) ¼"-internal diameter (ID) × ⁷⁄₁₆"-outer diameter (OD) × ³⁄₃₂"-wide rubber O-rings

(1) Clear polycarbonate or acrylic rod ½" in diameter × 12" in length

(1) Clear polycarbonate or acrylic rod ½" in diameter × 1¼" in length

(1) Clear polycarbonate or acrylic tube, ½" ID × ⅝" OD, 9½" long

(1) Ball knob, with a ½"-13 (½" in diameter, 13 threads to the inch) female thread × ⅝" depth, and a head that has a 1⅝" diameter (or, alternatively, you can glue a PVC tee fitting, ½"×½"×½", as a substitute for the threaded knob.)

Cyanoacrylate glue

Petroleum jelly

Epoxy glue

Plastic polish and steel wool

Sandpaper

Tools

Table saw, flat file, or lathe

½"-13 die and die handle

Electric drill and ¼-inch drill bit

Building the Fire Piston

Several steps in the building of the fire piston require careful attention to detail. Work carefully, especially when cutting the groove in the piston and fitting the O-ring into the groove.

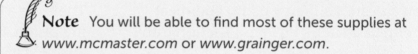

Note You will be able to find most of these supplies at *www.mcmaster.com* or *www.grainger.com*.

1. Refer to Figure 1.1. Cut the groove for the O-ring about ¼ of an inch from the end of the rod.

    The depth of the groove should be just slightly less than the diameter of the O-ring. If the groove is deeper, the O-ring won't seal against the tube properly. If the groove is too shallow, you won't be able to insert the rod into the tube.

    The best way to cut the groove is with a lathe. But if you don't have one, then what? Improvise! I used a table saw to cut the groove, and after a bit of trial and error, it worked fine.

    a. Raise the blade so the height of saw blade is protruding over the table an amount almost but not quite equal to the diameter of your O-ring.

    b. Carefully spin the rod as it contacts the blade to make an even slot (see Figure 1.2).

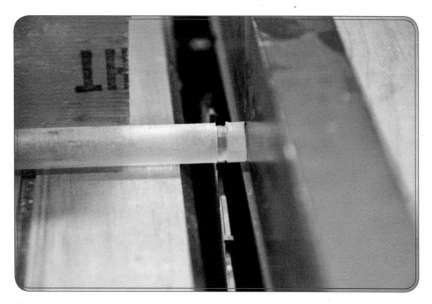

Figure 1.2: Cutting a groove

You might not be successful on your initial tries. Never fear, the rod is long enough so you can cut off mistakes and try again.

If you don't have a table saw or lathe, then you can carefully file the groove using a flat file that is almost but not quite the same thickness as the O-ring. If you're going to use a file, then select O-rings that are just a bit smaller in thickness than the thickness of the file.

Of course, it will take a fair amount of time to file away enough plastic to fit the O-ring, but with patience, it can be done. No matter what tool you use, work carefully because the depth and width of the groove must be quite accurate in order to seal the piston inside the cylinder.

2. Using the cyanoacrylate- or methylene chloride-based adhesive, glue the short rod into an open end of the tube as shown in Figure 1.3.

   It is very important that the glue makes the end air tight. Rotate the plug in the tube to distribute the glue.

3. Use a die and handle to cut a ½-inch-13 thread on the other end of the rod (see Figure 1.4).

4. Screw the ball knob onto the thread.

5. Insert the rod in the polycarbonate or acrylic tube and check the sliding fit.

Figure 1.3: Closing off one end of the tube by gluing a plug inside

Figure 1.4: Cutting the thread for the ball knob

6. Use a sanding belt, sandpaper, steel wool, and polish to make a close but free-sliding fit between the rod and the tube (see Figure 1.5).

Figure 1.5: Sanding the rod so it slides smoothly inside the cylinder

7. Cut the rod to a length of about 9½ inches. Optimally, there should be about ⅛ of an inch of space separating the end of the rod from the plug when the piston is fully inserted into the cylinder.

8. Install the O-ring into the slot you made in the rod in step 1. Depending on the width of the cut you made and the shape of the O-rings you procured, either one or two O-rings will fit in the slot (see Figure 1.6).

Figure 1.6: Rod with O-ring installed

9. Drill a ¼-inch hole (the divot), about ⅛-inch deep in the middle of one end of rod as shown in Figure 1.7.

The components of your fire piston are now complete (see Figure 1.8).

Figure 1.7: Drilling the divot

Figure 1.8: The completed fire piston

Using the Fire Piston

Now that you've assembled the parts of your fire piston, it's time to test it out.

1. First you need to test for air leaks. To do so, smear petroleum jelly on the O-ring and carefully insert the rod into the tube, working the O-ring past the edge of the tube.

    If you've done everything correctly, the piston will smoothly and easily pop back up, nearly to the top, when you release the knob after pressing down. If you press on the piston and it just stays in the tube, this means it won't work. If this happens you can troubleshoot with the following steps:

    a. Check for leaks in the plug by spraying the end with soapy water and looking for bubbles when you compress the piston.

    b. Improve the sliding fit by adjusting the depth of the O-ring groove and repolishing the surface of the piston.

2. Once you've got your piston moving smoothly, it's time to load it. Place a pinch of material that catches fire very easily in the divot you drilled in the end of the rod.

    The best material for this is called *charcloth*. I describe how to make charcloth later in this chapter.

3. Smear more petroleum jelly on the O-ring.

4. Carefully insert the rod into the piston, working the O-ring carefully past the edge of the tube.

5. Place the plug end of the fire piston on a hard surface.

6. Quickly and firmly press down on the knob. You'll see a bright flash in the bottom of the fire piston (Figure 1.9).

Figure 1.9: The tiny red glow means the charcloth has ignited.

7. Carefully remove the piston from the tube and blow on the glowing charcloth in the divot. You can now use the smoldering ember to start a larger fire.

How Fire Pistons Work

Unlike typical gasoline engines, the eponymous and now ubiquitous diesel engine has no spark plug or carburetor. Instead, the diesel engine works by compressing fuel under very high pressures. When the fuel/air mixture in the cylinder compresses, it also gets very hot. In fact, the mixture

quickly exceeds the flash temperature of the fuel and ignites.
The compressed gas expands violently upon ignition and
pushes the compressing piston away with enough force to
easily turn a drive train. Figure 1.10 explains this process in
more detail using a fire piston as an example.

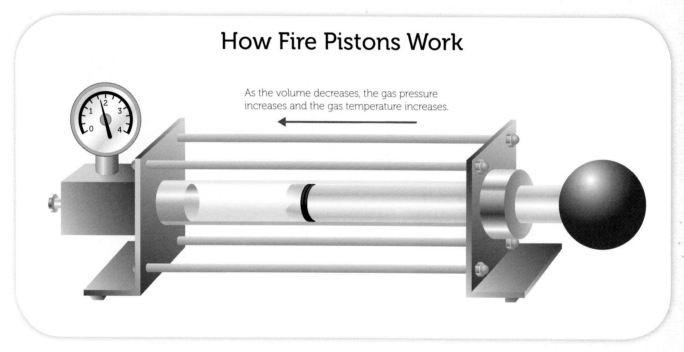

Inside the syringe of a fire piston, the air is trapped and cannot get by the rubber seals on the plunger. When you decrease the volume of space inside the piston by pushing the plunger, the pressure inside increases proportionally to the decrease in volume. This phenomenon is described mathematically by Boyle's Law.

At the same time, there's another physics law called Gay-Lussac's Law (we met Joseph Gay-Lussac earlier, in Volume 2 of this book series) that says when the pressure exerted on a volume of gas increases, the gas temperature increases proportionally.

Now if the pressure increase occurs slowly, then there's plenty of time for the heat to dissipate through the sides of the vessel holding it and nothing gets very hot. But if the pressure increase occurs very quickly, as in a fire piston, then the gas attains a high temperature and it's hot enough to start something on fire.

Figure 1.10: How fire pistons work

Making Charcloth

What is charcloth? Basically, it's cotton cloth that's been roasted to blackness at high temperature but in the absence of air. The wonderful thing about charcloth is that it is very easy to ignite with just a small spark. Charcloth doesn't really burst into flame easily, but it doesn't take much for it to catch fire and smolder, making it just right for starting something else on fire, like tinder or even a cigar.

To make charcloth, follow these steps:

1. Begin by taking an airtight metal container such as a candy tin and punching a small hole in the top (see Figure 1.11).

Figure 1.11: Make a small hole in a metal container.

2. Next, place some 100% cotton cloths in the tin and replace the top (see Figure 1.12).

3. Go outside and place the container on a handful of hot charcoal briquettes (see Figure 1.13).

   Almost immediately the cloth inside will start to roast and white smoke will pour out of the hole. After several minutes, the smoke volume will decrease or stop, signaling that the charcloth is done.

4. Remove the tin and let it cool. Once the container has cooled, you can remove the top and take out the charcloth.

Figure 1.12: Container with cotton cloth

Figure 1.13: Making the charcloth

Samuel Morse

Samuel Morse and the Telegraph

2

Prior to Samuel F. B. Morse's invention of the electric telegraph, there was no good way to communicate information across distance at any speed greater than that of a horserider. Sure, there were some limited experiments. Men holding flags called semaphores stood on hilltops and noted one another's coded movements through spyglasses. This was clumsy and slow. Worse, it was incredibly expensive because it required a lot of men on a lot of hilltops to relay messages over any substantial distance.

In the 1820s and early 1830s, Samuel F. B. Morse was well known as an accomplished painter; he was praised by artistic luminaries of the time: Gilbert Stuart, Benjamin West, and Washington Allston. In 1830, Morse left his home in New York for Europe, intending to refine his skills even further by studying with the masters of Italy, Switzerland, and France.

But while he was there, painting began to lose its hold over the ever-restless Morse. Although he was at the height of his artistic prowess, other ideas of how he might spend his life apparently began to encroach upon his thoughts.

From Artist to Inventor

In 1832, Morse boarded the packet boat Sully with his brushes and canvases and cruised westward across the Atlantic. While on board, Morse chanced upon a conversation with other passengers about the era's hot science topic—electromagnetism. What he heard fascinated him. More than that, this moment became a turning point not only in his life, but in the future of communications technology. Morse had boarded the ship an uninspired artist; he left it aspiring to be an inventor. As he left, he told the ship's captain, "Well, Captain Pell, should you ever hear of the telegraph one of these days as the wonder of the world, remember the discovery was made on the good ship Sully."

Despite the fact that he had little, if any, knowledge of electricity prior to this point, he plunged ahead as only a man in the throes of a serious midlife career crisis can. His lack of education and preparation in this field quickly became evident. In his crude laboratory, his homemade batteries had little or no energy, his elementary attempts at electromagnets produced no magnetism, and he didn't even understand the role of insulation on wires.

But Morse did have two important assets. He had plenty of determination, and he had smart friends. By asking the right questions of his friend Leonard Gale, a chemistry professor at New York University, he was able to make headway, building a telegraph capable of sending messages a quarter mile. A few years later, he asked the right questions of Gale's friend Joseph Henry, who was perhaps the most important American scientist living in the period between Benjamin Franklin and Thomas Edison. Armed with Henry's answers, Morse was able to design a workable long-distance telegraphy system.

In 1843, Morse received the then huge sum of $30,000 from the federal government to construct the first long-distance telegraph line, stretching from Washington, D.C., to Baltimore, MD, along the railroad tracks belonging to the B&O Railroad. The wires were strung from poles and trees along the right-of-way, insulated from the ground by crude insulators made from broken glass bottles. On May 24, 1844, the first telegraph message, "WHAT HATH GOD WROUGHT," was tapped from the B&O's Mount Clare Station in Baltimore to the Capitol Building in Washington. The era of electronic, high-speed information transfer had begun.

Making a Morse-Style Telegraph

You can make a telegraph system to communicate with friends, to practice Morse code, or simply to get a better appreciation for Samuel Morse and his world-changing invention. A telegraph system similar to the one Morse demonstrated in 1844 consists of four parts: the key, the sounder, the battery, and the wire.

In telegraphy, the key is a spring-loaded lever. When the telegraph operator presses the key, it completes an electrical circuit to the sounder, which is an electromagnet that pulls a steel plunger against a hard surface to make an audible click. The battery supplies the electrical voltage for the electromagnet, and the wire connects all of these items together.

Materials

(1) ¾"×½" aluminum angle iron, ¹⁄₁₆" thick, 5¾" in length

(1) ½"×½" square wood dowel piece, 5¾" in length

(1) ⁵⁄₁₆" round wooden dowel, ½" in length

(1) ⅜"-diameter light coil spring, ¾" in length

(2) ¾"×¾" corner braces

(1) ¾"×2" steel mending plate

(1) #8 machine screw, 1¼" in length, wing nut, washer

(2) #8 flat head wood screws, ½" in length

(2) #8 round head screws, ½" in length

(1) 1×4 wooden board, 7" in length

(1) Flat-shaped brass cabinet knob with machine screw

All-purpose glue

Tools

Variable-speed electric drill and drill bits

Screwdriver

Making the Key

Telegraph Key Directions

Follow these steps to construct the telegraph key:

1. To begin, you'll need to assemble the key's lever or tapper. To do so, line up the aluminum angle iron with the ½-inch square dowel along its length and then glue it to the dowel shown in Figure 2.1.

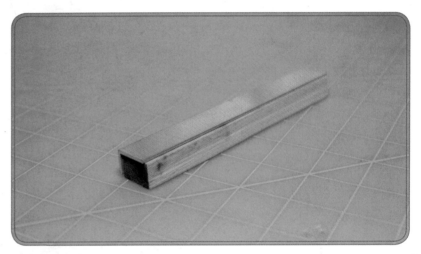

Figure 2.1: The aluminum angle iron glued to the dowel

2. When the glue is dry, drill a centered ³⁄₁₆-inch hole through the lever/tapper for the knob screw ¼ of an inch from the end of the dowel (usually the knob screw is a #8 machine screw, but check your knob before you drill) as shown in Figure 2.2.

Figure 2.2: Drilling the hole for the knob screw ·

3. Next, drill a centered ⅛-inch hole in the side of the lever/tapper, 2 inches from the same end of the dowel as the knob screw hole, as shown in Figure 2.3.

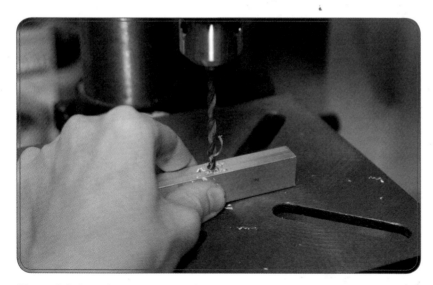

Figure 2.3: Hole in the telegraph's tapper

4. Attach the tapper knob as shown in Figure 2.4.

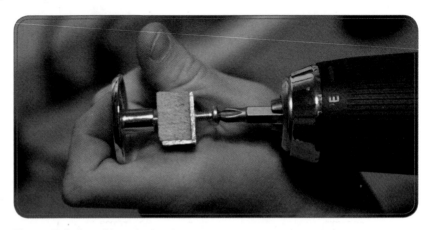

Figure 2.4: Attaching the knob

This completes the lever/tapper. Set this aside for now so you can begin working on the telegraph key base.

5. Refer to Figure 2.5, which shows the telegraph key's base, for a visual and drill the following holes.

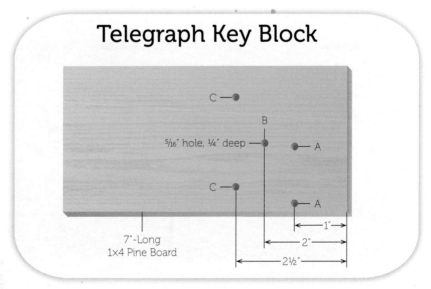

Figure 2.5: Top view of the telegraph block (just holes)

a. As shown in Figure 2.5A, drill a centered ⁵⁄₁₆-inch-diameter hole (labeled "B" in Figure 2.5) 2 inches from the edge of the board for the dowel that will hold the spring.

Glue the ⁵⁄₁₆-inch round wooden dowel into the hole. The dowel end should extend about ¼ inch above the surface of the block. When the glue is dry, place the spring over the dowel as shown in Figure 2.5B.

b. Using the steel mending plate as a guide, drill two ⅛-inch pilot holes (labeled "A" in Figure 2.5) for the #8 round head wood screws 1 inch from the edge of the 7-inch-long 1×4 wooden board.

c. As shown in Figure 2.5C, drill two more ⅛-inch pilot holes (labeled "C" in Figure 2.5) 2½ inches from the edge of the 1×4 for the corner braces. The holes should be placed so the vertical sections of the corner braces are 1 inch apart.

Figure 2.5A: Drilling a ⁵⁄₁₆-inch hole

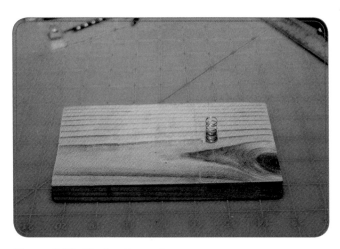

Figure 2.5B: Placing the spring

Figure 2.5C: Drilling two pilot holes

Making a Morse-Style Telegraph **25**

Figure 2.6: Mending plate attached

6. Now attach the 2-inch-long steel mending plate to the telegraph key base with the round head screws using the pilot holes you drilled in step 4 (see Figure 2.6).

7. Then attach the ¾-inch by ¾-inch corner braces to the wood block with the #8 flat head wood screws using the pilot holes you drilled in step 4c.

   The partially assembled telegraph key should look like Figure 2.7.

8. To assemble the two parts of the telegraph key block (the tapper and the base), insert the 1¼-inch #8 machine screw through the hole in one of the ¾-inch corner braces, into the hole you drilled through the tapper in step 3, and then into the other corner brace.

Figure 2.7: The partially assembled telegraph key

9. When you've got everything lined up, attach the wing nut.

When you're done, your telegraph key should look like the one in Figure 2.8.

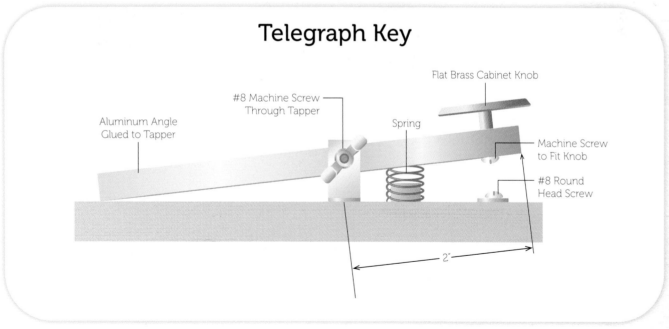

Figure 2.8: Completed telegraph key

Materials:

45′ of 26-gauge coated magnet wire

(1) 5⁄16″ bolt, 2″ in length, matching nut and washer

(1) 12-gauge steel strip, 1″×12″

(2) 12-gauge steel strips, 1″×3½″

(4) #8 wood screws, ½″ in length, washers

(1) Wood 2×4, 6½″ in length

(1) 1×4 wooden board, 12″ in length

(2) 2″ nails or deck screws

Making the Sounder

The next step to assembling your telegraph system is to construct the sounder.

Sounder Directions

Follow these steps to assemble your sounder:

1. Drill holes on the faces of the wooden boards and metal strips as indicated in Figures 2.9 and 2.10.

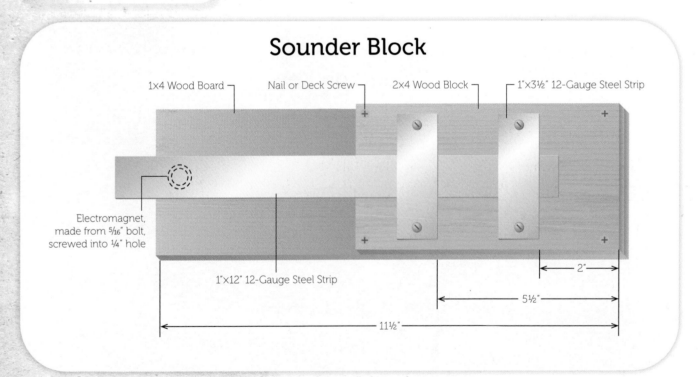

Sounder Block

1×4 Wood Board   Nail or Deck Screw   2×4 Wood Block   1″×3½″ 12-Gauge Steel Strip

Electromagnet, made from 5⁄16″ bolt, screwed into ¼″ hole

1″×12″ 12-Gauge Steel Strip

2″

5½″

11½″

Figure 2.9: Arial view of the sounder block assembly

2. Fasten the 1×4 and 2×4 boards together with their edges lined up using nails or deck screws, as shown in Figure 2.11.

Tools:

Variable-speed electric drill and drill bits

Screwdriver/hammer

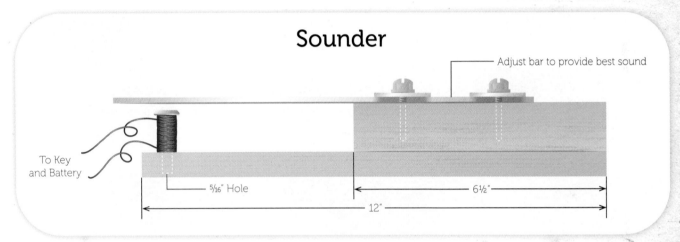

Sounder

Adjust bar to provide best sound

To Key and Battery

⁵⁄₁₆" Hole

6½"

12"

Figure 2.10: Side view of the sounder block assembly

Figure 2.11: Joining the 1×4 and 2×4 boards

3. Place the 5/16-inch bolt in the chuck of your drill so that 1 inch of screw thread is exposed.

4. Leaving about 1 foot of wire trailing, wrap the magnet wire around the threads closest to the head of the drill.

5. With a helper holding the magnet wire spool so it is free to spin, start the drill and wrap the wire around the bolt, using an up and down motion so the wire turns are distributed as evenly as possible. Stop winding when there is a foot of wire left and tie it off so it does not unravel (see Figures 2.12 and 2.13).

Figure 2.12: Wrapping the magnet wire around the bolt to make your electromagnet

Figure 2.13: A close-up of the wrapped wire

6. Screw the electromagnet into the 5/16-inch hole you drilled in the 1×4 in step 1 to a depth of about ¼ inch. (Since the screw is much harder than the wood, you should be able to screw the bolt in fairly easily without needing to tap the hole first.)

7. Screw both of the 3½-inch-long steel strips onto the 2×4, as shown in Figure 2.9, but don't screw them tight— leave enough space for the 12-inch metal strip to slide underneath.

8. Insert the 12-inch-long steel strip between the 3½-inch metal strips and the 2×4. Center the long strip on the 2×4 and tighten the screws attached to the 3½-inch-long steel strips slightly.

The completed sounder look should like Figure 2.14.

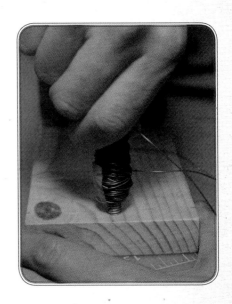

Figure 2.14: The completed sounder

Wiring the Telegraph

Now that you've completed the assembly of both the key and the sounder, it's time to wire your telegraph system (as shown in Figures 2.15 and 2.16.

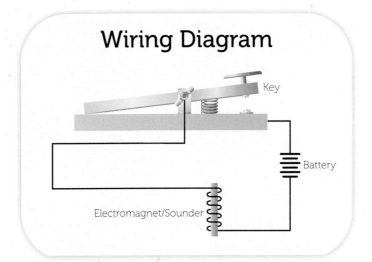

Figure 2.15: A diagram showing how to wire the key and sounder

Figure 2.16: This photo shows what the system looks like after it is wired.

After you've wired your system, adjust the placement of the 12-inch strip on the sounder so it responds quickly and precisely when the telegraph key is closed. When you have it well adjusted, tighten down the cross strips to hold it fast.

How the Telegraph Works

Basically, the telegraph is a pair of electromagnets attached to spring-operated switches. As Figure 2.17 shows, when the key is pressed, the circuit is completed, causing the sounder electromagnet on the distant end to energize and make a click sound. When the key is released, the sounder springs back up.

It takes a fair amount of electrical energy to operate long-distance telegraph electromagnets because the telegraph wire does have some electrical resistance, and over long distances, that adds up. One of the important ways Morse made the telegraph work over long distances was by incorporating Joseph Henry's electromagnetic relay, which allowed for the use of large storage batteries along the telegraph line.

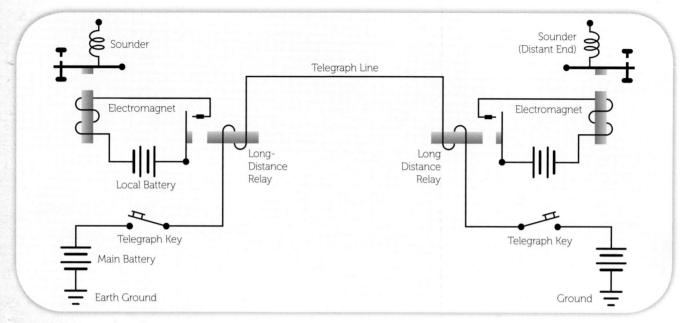

Morse's telegraph allowed communications between distant points. To communicate, the telegrapher pressed down on a telegraph key, which was a simple electrical switch. When the key was pressed, the circuit was completed, allowing electricity to flow, and when it was released, the electricity stopped flowing.

When the electricity flowed, the electromagnet energized, causing the sounder to make an audible click, both in the local telegraph office and in the distant office. The distant office was connected through a series of relays, batteries, and long-distance wires.

There were two batteries in the circuit: one for activating the local sounder, which was called the local battery, and one for sending the circuit out through the telegraph wires to the distant end, which was called the main battery. The local battery was small, with a voltage of around 1.5 volts (like a modern D-cell). The voltage of the main battery was much larger, perhaps 100 volts or more, depending on how long the telegraph wire was.

Figure 2.17: Explanation of how the telegraph works

A *relay* is a switch that opens or closes an electrical circuit based on input from another circuit. That means a small, nimble switch like a telegraph key can, with the use of a relay, control a big, long-distance telegraph circuit with the sounder located hundreds of miles away.

With multiple relays, electrical signals can travel any distance and arrive at the receiving station with nearly as much power as they had when they were transmitted.

Morse Code

Since the telegraph transmits clicks, not speech, Samuel Morse had to invent a system or code in which alphabet letters and numbers are represented by combinations of long and short sound telegraph signals. Figure 2.18 shows the long and short of Morse code.

Study Figure 2.18; once you learn the code, you can understand the following joke. The periods represent dots and the hyphens are dashes; the slashes separate the words in the sentence; and the second "letter" (---...) is a colon.

--.- ---... / .-- .... .- - / -.. --- / -.-- --- ..- / --.. - / .. ..-. / -.-- --- ..- / -.-. .-. .-. --- .... / .- / .--. .- .-. .-. --- - / .-- .. - .... / .- / .-- --- --- -.. .--. . .-. -.- .-. . .-. . --- . .--.. / .- ---... / .- / -... .. -. -.. / - ..... .- - /
- .- .-.. -.- ... / .. -. / -- --- .-. ... . / -.-. --- -...

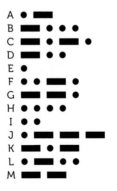

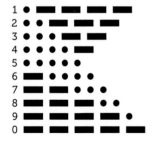

International Morse Code

1. The length of a dot is one unit.
2. A dash is three units.
3. The space between parts of the same letter is one unit.
4. The space between letters is three units.
5. The space between words is seven units.

Figure 2.18: Morse code

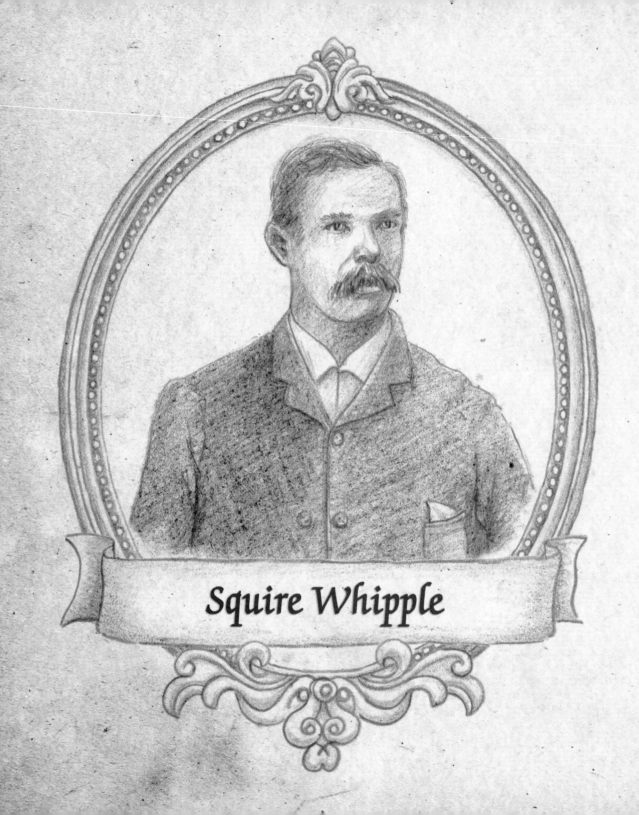

Squire Whipple

Squire Whipple and the Truss Bridge

3

All mechanical, civil, and aerospace engineers learn how to analyze structures early in their engineering training. Determining the size and direction of compressive or tension forces acting on each piece of a large construction is a civil engineering discipline called *statics*. With this basic knowledge, engineers figure out the right sizes, shapes, and thicknesses of materials they use to build houses, dams, bridges, buildings, and other nonmoving structures.

When studying statics, engineering students learn about many architectural elements that are designed to hold things up against the forces of gravity—arches, beams, buttresses, and vaults, to name a few. The *truss* is one of the most important of the basic building elements and it's widely used in construction. In its simplest form, a truss is merely a rigid framework of bolted-together triangles. Triangles are inherently strong and stable, and structures made out of them are also strong, rigid, and lightweight.

The Origin of the Truss

Who first figured out how to build with trusses? It's hard to say. Certainly, the practice has been around for a long time, but the archeological records are too sketchy to tell us exactly who came up with the idea. It's clear that the classical Greeks, for all their genius in other fields, didn't know much, if anything, about building with triangle trusses. The Romans may have dabbled with them, but examples of Roman buildings and viaducts that use them are few and far between.

But something must have happened during the early period of the Dark Ages that followed the fall of the Roman Empire, because later in Medieval times, the truss was well known to builders. Medieval cathedral and church architects empirically, if not scientifically, understood the technique of truss building, and there are plenty of examples of early European buildings in which wooden triangular trusses hold up the roof.

However, really understanding the physics behind the truss took a long while to come about. Engineering practice in both the Middle Ages and Enlightenment was not analytical or based on scientific knowledge as much as it was empirical and experience based. Today, structures like bridges are designed using science. Knowing how to design a truss scientifically is an important skill for which engineers can thank New York civil engineer Squire Whipple (Squire was his first name, not his title), who developed the first method for analyzing and designing trusses.

In 1841, Whipple published *A Work on Bridge Building*, which revolutionized the field of civil engineering. No longer would builders use "rules of thumb" to guess at how big and thick to make a strut or girder. Now, because of Whipple's work, they knew *exactly*.

Whipple and the Design of Structures

Prior to the publication of this book in 1841, Whipple was busy. During this time, he figured out how to analyze the forces acting on a stationary object using a graphical method of lining up, head-to-toe, representations of the size and directions of forces that he called *the polygon of forces*. Nothing of this method depended on complex math, just the clever use of rulers and protractors on graph paper.

A structure such as a bridge is subject to many different simultaneous forces. At any chosen point on the structure, each of those forces can be represented by an arrow of a fixed length and direction. Whipple determined that you can take all those force arrows, called vectors, and line them up, head to toe, to make an open-sided polygon. The vector required to close the polygon is called the *resultant*, and the length and direction of the resultant is the net force on the structure at that location.

To make his method work, Whipple made a number of assumptions. Most important is the assumption that all connections are pinned rather than fixed and are free to rotate. (This is not really how bridges are typically built. Normally, steel gusset plates are used to connect one member to another, and they are solidly riveted, so no rotation is possible. But this assumption makes little difference to the accuracy of calculations in most cases, and it greatly simplifies matters by ensuring that all forces on all members are applied parallel to the centerline of the member.)

Whipple designed a number of bridges during his career. One of his designs, a bowstring-shaped truss bridge made of cast iron, became the standard design for bridges over the Erie Canal. For his contributions, the Society of American Civil Engineers declared Whipple the "father of American iron bridge building."

The publication of Whipple's book set off a boom in civil engineering projects. Because this concept wasn't so hard to understand and didn't require them to know complex math like calculus or trigonometry, engineers quickly picked up on this knowledge, and with the polygon of forces at their disposal, they began to safely and economically design viable truss bridges out of wood and steel members.

Soon after this publication, the great age of iron bridge building was in full swing. The triangle structure of the truss is obvious in the beautiful spans of the Navajo Bridge over the Colorado River in northern Arizona, the Whirlpool Rapids Bridge over Niagara Falls, and the Quebec Bridge over the St. Lawrence River. But hidden trusses are also important parts of suspension and cantilevered bridges such as the Brooklyn Bridge in New York (see Figure 3.1).

Brooklyn Bridge

Quebec Bridge

Whirlpool Bridge

Navajo Bridge

Figure 3.1: Examples of truss bridges

Building a Warren Truss Bridge

There are many different types of truss bridges, and each type is typically named after the engineer who first devised it, but they all were influenced by Whipple's work. The simplest bridges are the Howe, the Pratt, and the Warren, all shown in Figure 3.2.

Materials

100 wooden craft sticks (6"×11/16"—tongue-depressor-sized)

(2) Standard bricks (2¼"×3¾"×8")

8"×8" sheet of ⅛"-thick balsa wood

Tools

Utility knife

Masking tape

Ruler

Hot glue gun and glue

Plate weights or concrete blocks for testing structure (optional)

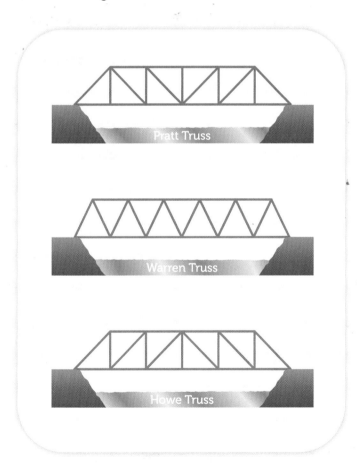

Figure 3.2: Truss types

In 1848, English engineer James Warren designed the first truss bridge that consisted solely of equilateral triangles connected side by side. The design starts with a single triangle; the next triangle is inverted and connected to the first. The following triangle is again inverted and connected, and the structure continues in this manner across the span of water. Structural engineering doesn't get much simpler than this, but the design of the Warren truss is also tried and true—it produces a bridge that is easy to build, strong, and relatively lightweight.

In this project, we'll make a model Warren truss bridge. Light and strong, our bridge model can hold more than 100 pounds. If you build it very carefully, it can hold a lot more than that!

Constructing the Bridge

Now that you've assembled your supplies, it's time to begin:

1. Use the utility knife to cut the balsa wood into 14 squares measuring 2 inches on a side. These squares are the gusset plates for the bridge (see Figure 3.3).

2. Using the craft sticks, gusset plates, glue gun, and glue, you'll now build the two bridge sides. For the first side, using the masking tape, tape the gusset plates to your work surface. Then hot glue the craft sticks to the gussets as shown in Figure 3.4.

   Take care to make the glued connections neat, and make sure you align the craft sticks to form exact equilateral triangles.

Figure 3.3: The gusset plates

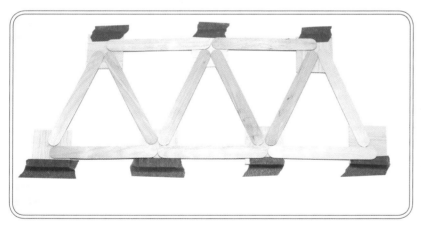

Figure 3.4: Building the first of two bridge sides

3. While the glue sets on the first bridge side, glue together the second side the same way.

4. Once the glue sets on the first side, flip it over, and attach craft sticks to the other side in the same positions as you did with the first side.

5. Do the same with the second truss (see Figure 3.5).

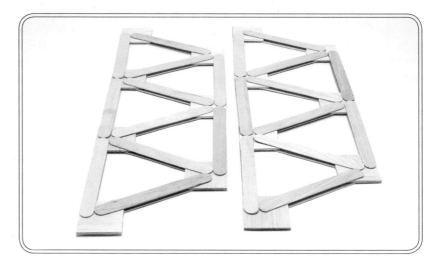

Figure 3.5: The completed pair of trusses

The long top and bottom members of the sides are called *chords*. The slanted pieces that tie the chords together are called *struts* or *braces*.

6. If you use Squire Whipple's polygon of forces method to analyze the forces in each member, you'll find that the greatest forces are in the top and bottom chords, at the center of the bridge. So, to make your bridge its strongest, glue on an extra craft stick at those points to reinforce both trusses.

Now that you've got your trusses built, it's time to move on to build the rest of your bridge structure.

7. Place the bricks on your work surface 4 inches apart and make certain that their long sides are parallel.

8. Tape each truss to one of the bricks.

9. Using more craft sticks, glue the struts and braces across the top and bottom and ends of the bridge as shown in Figure 3.6.

Figure 3.6: Gluing the two trusses together

When you are finished connecting the two sides, your bridge should look similar to Figure 3.7.

> **Warning:** Making sure that the sides are exactly vertical is essential. Even a small amount of leaning will cause your bridge to fail prematurely. The bridge's sides must be perfectly perpendicular to the horizontal members that connect the two trusses.

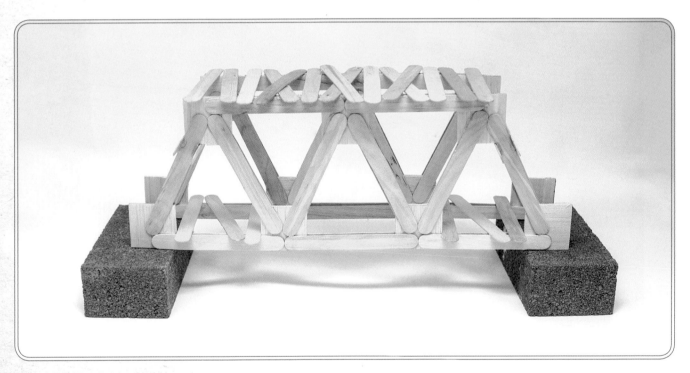

Figure 3.7: The finished bridge

Testing Your Truss Bridge

Now that you've got your bridge constructed, it's time to see exactly how strong it is! Follow these steps to find out.

1. Place the bricks 14 inches apart on the floor.

2. Place the front and back edges of the bridge on the bricks (as in Figure 3.8).

3. Load 'er up! You can use concrete blocks, barbell weights, buckets of water, or anything else flat and heavy (see Figures 3.8 and 3.9).

   Add weight slowly and incrementally and make sure you keep body parts like toes out from underneath the bridge. Keeping the weight evenly distributed will allow you to add more weight before the structure fails.

Figure 3.8: Bridge with load

Figure 3.9: Bridge with heavy load

It's your bridge, so you can either add weight until it eventually fails, or you can paint it and display it proudly as proof of your engineering abilities.

How Truss Bridges Stand

The reason engineers design structures with trusses is that they efficiently and inexpensively transfer weight from long, horizontal members to strong vertical ones. For example, a truss bridge transfers the weight of the cars driving over it to the bridge's study pilings and piers. Similarly, the trusses holding up the roof of your house basically take the weight of the roof's shingles and rafters and move it to the strong vertical posts and studs inside the walls. Because of the triangle-based geometry of the truss, they can be lightweight and relatively inexpensively. Figure 3.10 explains this in more detail.

If you look carefully at a Warren, Platt, or Howe truss bridge, what do you see?

You see a bunch of triangles! That's because triangles, unlike rectangles or other polygons, are naturally strong and stable structures. Triangles have three sides and no matter how you attach the end of one side to another side, a triangle is the only configuration it can take. If you press hard on a triangle, even if you don't torque down the bolts connecting the sides, it's still a triangle.

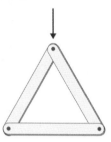

A rectangle, on the other hand, can take the form of a square or a rhombus, or a diamond shape. That means rectangles can be squashed and are not nearly so rigid. Sure, you can tighten the bolts holding the rectangle, but eventually it will still skew or squish if you put a load on it.

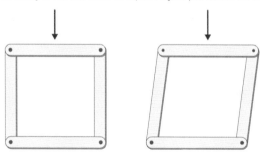

So if you want to make a structure that's rigid, regardless of the rigidity of its connections, you start with a triangle and build onto it by adding sticks to make more triangles. That, in a nutshell, is a truss.

Figure 3.10: How trusses work

Lee de Forest

Lee de Forest and the Amplifier

4

In this chapter we'll delve into the early history of electronics. In Chapter 2 we met Samuel Morse and explored the science behind his telegraph. Morse's telegraph is an electrical device. In this chapter we'll look at electronic devices. Let's begin by understanding the difference between an electrical device and an electronic one.

You likely already understand, perhaps intuitively, that light bulbs, space heaters, and kitchen mixers are electrical devices whereas computers, televisions, and radios are electronic ones. But what is the scientific difference?

It comes down to the way electricity is used inside the device. *Electrical devices* use electric current and change it into another form of energy. For example, electric light bulbs take electric current and turn it into heat and light; your electric kitchen mixer converts electricity into the spinning motion of the beaters.

In contrast, *electronic devices* don't simply convert electrical current into other forms of energy; instead they manipulate the electrical current itself to do useful things. Electronic devices actually change the electrical currents into different shapes and frequencies. Electronic devices shape electricity into music, as in a radio, or into visual images, as in television.

De Forest Develops the Amplifying Vacuum Tube

Because he was the son of an Alabama minister, everyone expected that young Lee de Forest would follow in his dad's footsteps, living a spiritual life and preaching to his congregations. But, it turned out that de Forest was fascinated by science, not theology. He possessed the important gifts that an inventor needs—not only was he good with tools, but he had a passion for making new things. In 1899, he got his PhD from Yale and started working for the Western Electric Company in Chicago. He did well, but before long, he left in order to work independently on projects of his own choosing.

De Forest's greatest achievement came in 1906 when he developed the first amplifying vacuum tube. This tube could do something new and extremely important: it could increase the power of radio signals.

Up until that point; early radio had a big problem: the signals received by radio sets were so weak that they could only be heard through headphones. Scientists knew that if radio was ever to become really popular, the sound it made had to be loud enough for several people to hear it at once, which meant it needed a loudspeaker (see Chapter 8, "Chester Rice and the Loud-speaker"), and loudspeakers require strong, powerful electrical signals.

De Forest's invention, which was called the Audion, was an electronic device that could transform a small electrical signal coming in one side of the device into a much larger one when it exited the other side.

The Audion tube had three connections: the anode, the cathode, and the control grid. Current passing through the filament, or cathode, heated it up, which caused it to emit a stream of electrons. The electrons, being negatively charged, were attracted to the positive plate at the top of the Audion. De Forest's world-changing idea was to place a grid of wires between the filament and the positive plate. When this was done, the grid would become more or less neg-atively charged as more or less voltage was applied to it. Because the signal to the grid varied, it would control the number of electrons flowing between the filament and the anode. This was how a tube ampli-fier increased the size of the signal—it

used a small signal to control a much large voltage.

Once engineers figured out how useful amplifiers could be, they incorporated them into radios, telephone systems, scientific instruments, and much more. The vacuum-tube business exploded. During the next 50 years, millions of tubes were manufac-tured. But these tubes had a lot of short-comings—most significantly, they were big, expensive, and hot.

So, in 1945, Bell Labs put together a team of scientists to find a better alternative to the vacuum tube. The team came up with another sort of amplifier called the *transis-tor*. Like a vacuum tube, this silicon-based or solid-state amplifier also has three parts. Current applied to the middle part controlled a much larger current or voltage between the top and bottom parts.

Modern amplifiers are made with inte-grated circuits that contain the equivalent of thousands or even millions or transis-tors or vacuum tubes in a single electronic component.

Building an Amplifier-Based Touch Switch

In the following exercise, you'll use a couple of simple solid-state transistors (although, in principle, you could make de Forest's vacuum tube Audions work as well) to amplify a tiny bit of electrical current moving through your finger. Using transistors, you are able to construct a device that allows you to build a touch switch that completes a circuit to light light-emitting diodes (LEDs) or sound a buzzer.

Your body has a fair amount of electrical resistance, which is why you can't just grab two pieces of wire with your hands and complete a low-voltage (6 volts, direct current) circuit to make an LED light up. That's not to say that no current flows at all; it's just that your skin presents so much resistance (about 100,000 ohms, depending on how clean and dry your skin is) that the amount of amperage coursing through you is exceedingly small. But if that small amount of electrical flow could be detected and then increased, say through the use of an electronic amplifier, then you could use your finger alone as a switch. In this exercise, we'll use *electronic amplification*, the concept pioneered by Lee de Forest, to make a touch switch.

Before you get started, refer to Figure 4.1, which shows the circuit diagram you will build.

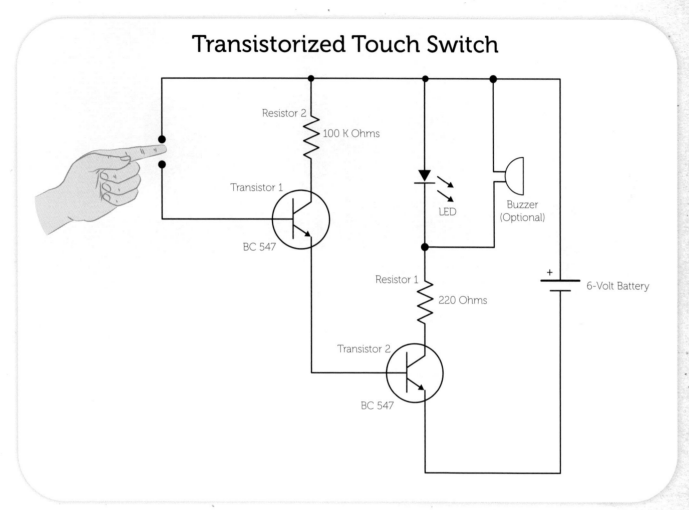

Figure 4.1: The touch switch

To assemble this circuit, complete the following steps.

1. Refer to Figure 4.2. Examine your LED and identify the positive lead (the longer one).

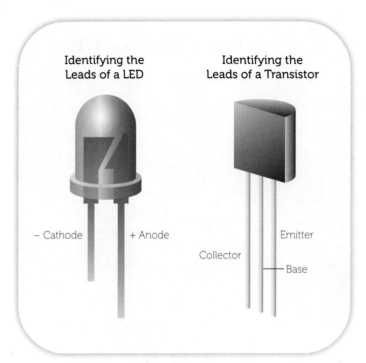

Figure 4.2: Learning to identify leads

2. Hold the transistor with its flat side toward you and its legs pointing down, and identify the collector (left leg), the base (center leg), and the emitter (the right leg).

Now it's time to wire your circuit. If you know how to use a prototyping board, then it's a very easy setup. If you don't, you can connect the components together in space, and then solder them.

3. Begin by wiring the transistors together so the emitter of the first-stage transistor is connected to the base of the second stage resistor.

4. Then connect the emitter of the second-stage resistor to the negative battery terminal as shown in Figure 4.3.

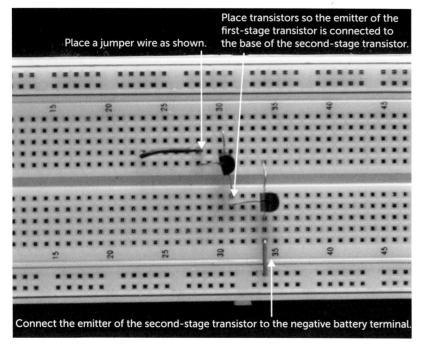

Figure 4.3: Wiring the transistors together

5. Place a 100 K ohm resistor between the first-stage transistor collector and the positive battery terminal.

6. Next, connect the positive end of the LED to the positive battery terminal.

7. Now insert a 220 ohm resistor between the negative LED terminal and the collector of the second-stage resistor as shown in Figure 4.4.

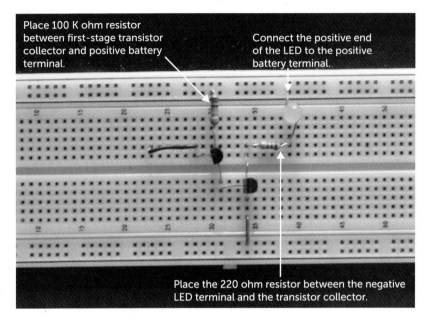

Place 100 K ohm resistor between first-stage transistor collector and positive battery terminal.

Connect the positive end of the LED to the positive battery terminal.

Place the 220 ohm resistor between the negative LED terminal and the transistor collector.

Figure 4.4: Adding a 220 ohm resistor

8. Now refer to Figure 4.5. You'll need to use two jumper wires placed vertically on the prototyping board as your touch switch. Place one of the wires on the positive 6-volt rail and connect the other to the first-tage transistor base.

9. Now connect a buzzer in parallel with the LED if you want to.

10. Finally, connect your 6-volt battery pack to the positive and negative rails of the prototyping board as shown in Figure 4.5.

In this circuit, the amplifiers are wired in a series to provide a gain, or amplification, that is several hundred times the original current. When you press on the bare wire connected to the 6-volt rail simultaneously with the base of the transistor connected to the 100 K ohm resistor, a tiny amount of current flows from the battery, through your finger, to the transistor base. But that small current is amplified first by

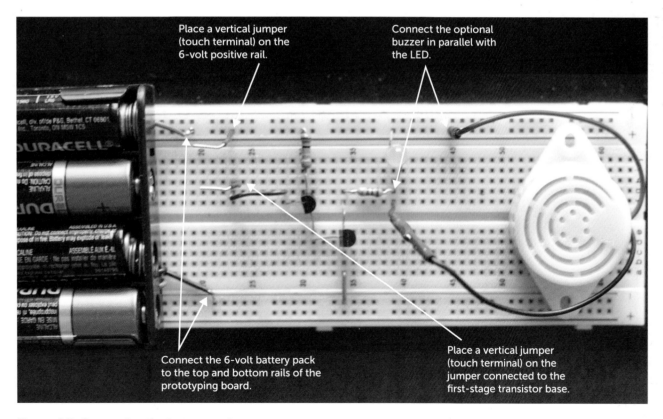

Place a vertical jumper (touch terminal) on the 6-volt positive rail.

Connect the optional buzzer in parallel with the LED.

Connect the 6-volt battery pack to the top and bottom rails of the prototyping board.

Place a vertical jumper (touch terminal) on the jumper connected to the first-stage transistor base.

Figure 4.5: Connecting the battery pack

one transistor, then the second, increasing it enough to light the LED. Congrats, you've made your own amplifying touch switch!

How Amplifiers Work

Amplifiers are devices that take a small signal (imagine a sine wave) and turn it into a larger signal (a much bigger sine wave). The amplified signal looks exactly like the original signal, just much larger. All amplifiers, whether they use vacuum tubes, transistors, or silicon chips, are designed to do just this.

The original de Forest vacuum tube consisted of a high-voltage positive and negative electrical terminal separated by an electrically conducting grid, all locked inside an evacuated glass tube. Inside the tube, a red-hot electric filament spewed a cloud of electrons that were attracted to the oppositely charged plate at the far end of the tube, but were moderated by the presence of an electrical grid (see Figure 4.6).

Unfortunately, tube amplifiers like these are inefficient because they use a great deal of power, break down quite a lot, and give off a great deal of heat.

The invention of the transistor in the late 1940s changed the world of electronics. The transistor uses semiconductors instead of heated terminals, and it uses a control grid to control currents and voltages. Since they are physically much smaller and consume significantly less power, electronic devices built with transistors are faster, cooler, and more reliable than those with tubes.

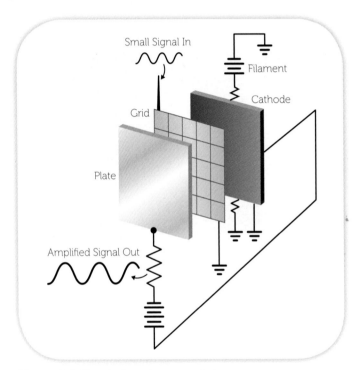

Figure 4.6: The de Forest vacuum tube

De Forest's vacuum tube is a special type of tube known as a triode.

Inside the glass wall of a triode, an electrically charged plate called a *cathode* is heated. The heat causes the electrons to leave the positively charged cathode terminal, much like steam molecules rising from a pot of boiling water. Where do the electrons go? They move toward the negatively charge terminal, which is usually just called the *plate*.

The "boiled off" electrons flow from the plate through the battery to the cathode and a current is established.

But in a triode, between the cathode and the plate is a grid of wires that are attached to a voltage source that changes in proportion to something else—for example, the small, changing amplitude of an audio signal. As the small signal changes, the grid becomes more or less positively charged, allowing more or fewer electrons to move from the cathode to the plate.

When a resistor is attached to the negative side of the triode, you'll find that the small changing voltage at the grid causes much larger proportional change in voltage between the cathode and plate.

That's the essence of electronic amplification—a small voltage change at one point in the circuit causes a much larger but proportional voltage change in another.

August Möbius

August Möbius and the Möbius Strip

5

In 1905, physician Paul Möbius included a photo of his long-dead grandfather's (the great mathematician August Ferdinand Möbius) skull juxtaposed with the skull of Ludwig von Beethoven in his new book, *Ueber den Schädel eines Mathematikers*. The German title translates to *Over the Skull of a Mathematician*. It's not completely clear how Paul obtained the photo of Grandpa Möbius' skull, much less that of Beethoven's, but there they are, captured in black and white.

This odd photo was made in the name of weird science. Paul Möbius was a neurologist of some distinction who dabbled in the bunk science of phrenology. He is thought to have compared the two skulls in an effort to prove that certain skull shapes were associated with mathematical and musical aptitude.

Although Möbius the Younger's head bump research never panned out, he was right about one thing: August Möbius, professor of astronomy and observatory director at the University of Leipzig, was unquestionably a man of incredible mathematical ability. August Möbius' scientific contributions are enormous: he invented a new type of calculus, he advanced the fields of celestial mechanics and astronomy, and he provided numerous important insights into the study of geometry.

But he is most famous now for his work in the realm of topology, the mathematical study of shape and form. Unfortunately, appreciating much of Möbius' highly complex and arcane nature work requires a familiarity with concepts such as geometrical mechanics, polyhedral boundary theory, projective transformations, and other eye-glazing, albeit no doubt important, mathematical concepts.

But there is one Möbian concept that everyone can appreciate and that is the wonderful Möbius strip.

The Arrival of the Möbius Strip

In 1858, while working on some incredibly complex topological concepts involving geometrical solids, Möbius took a band of thin, flat material, gave it a single twist, and fastened the ends together. The Möbius strip was born. (To be fair, it was also independently discovered by another mathematician named Johann Listing at nearly the same time, but nobody calls it a Listing strip.)

In topological terms, a Möbius strip is a three-dimensional surface with only one side, and it has some amazing physical properties. For example, if you draw a line starting from the seam down the middle of the strip, the pencil line will meet back at the seam but on the "other side." Also, if you cut a Möbius strip along the center line, you get not two separate strips, but rather one long strip with two complete axial twists. If you attach a Möbius strip to an object, say a bowling pin, and swing it around your head, the twisted strip resists kinking and curling, proving itself to be a superior attachment to a non-twisted one.

In the following activity, we exploit the properties of Dr. Möbius' strip to make the Möbius strip coffee cup carrier, a Möbius strip–equipped cup holder that's nearly spill-proof.

The Möbius Strip Coffee Cup Carrier

Every time you drive through a highway entrance cloverleaf, you and the passengers in your car experience centrifugal force, which pushes you toward one side of the car. Another common example of this is the way the clothes in your washing machine press against the rotating drum during the spin cycle.

For such a common occurrence, centrifugal force is a bit hard to explain. Physics books define it as "the apparent force, equal and opposite to the centripetal force, drawing a rotating body away from the center of rotation, caused by the inertia of the body." Maybe that helps (probably not, though), but at least we can say with certainty that when something gets spun around in a big circle, centrifugal force causes it to press outward, opposite the central point of rotation. We'll use this phenomenon and the anti-kinking abilities of the Möbius strip to design an anti-spill coffee cup holder.

Materials

(1) Piece of 12"×12" acrylic plastic, ⅛" thick

(2) Split rings, 1" in diameter

(1) Large barrel swivel (available where fishing tackle is sold)

(1) 3 ½"×3 ½" piece of leather

Soft leather strip, ½"×8"

Acrylic adhesive/solvent

Glue

Tools

Jigsaw/table saw

Hot air gun (for paint stripping)

Electric drill with ⅛" drill bit

Metal vise

(2) Medium clamps

Sandpaper

Pop rivet gun and rivets, or stout needle and thread

Ruler

Marking pencil

Making the Cup Holder

Follow these steps to make the cup holder:

1. Cut a 3½-inch-wide strip from the acrylic plastic square with the jigsaw (or table saw if you've got one) as shown in Figure 5.1.

Figure 5.1: Cutting a strip

2. Measure your favorite coffee cup and then make appropriate bend lines on the 3½×12-inch piece of acrylic plastic as shown in Figure 5.2.

Figure 5.2: Bend lines marked

3. Place the marked plastic piece in the vise and direct the hot air from the air gun at the bend line.

4. When the plastic softens, bend the piece 90 degrees at the first bend line.

5. Make additional bends so the plastic takes the shape shown in Figure 5.3.

> **Warning:** Use care to avoid burns. The hot air gun and plastic can become extremely hot!

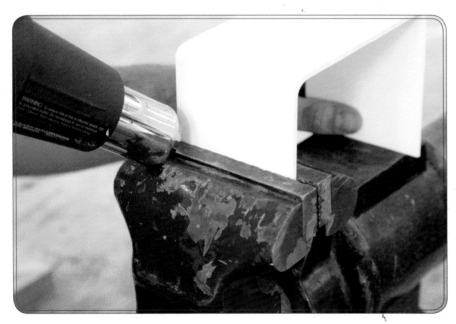

Figure 5.3: Bending the plastic

6. Using the bent plastic piece as a form, trace out and mark side pieces that are nominally 6 inches by 1½ inches (see Figure 5.4).

Figure 5.4: Marking side pieces for the cup holder

Make sure to match the side contour of the main 12-inch piece of plastic as closely as possible when you are marking the plastic.

7. Cut out the marked pieces with the jigsaw (see Figure 5.5).

8. Now clamp the sides to the bent main piece (see Figure 5.6). Using acrylic adhesive, glue the side pieces to the main piece (see Figure 5.7).

Figure 5.5: Cutting the side pieces

Figure 5.6: Clamp the sides in place and let the adhesive set

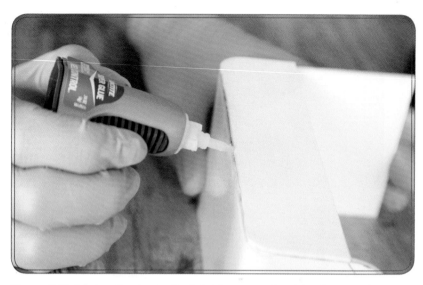

Figure 5.7: Apply adhesive to the joint between the side pieces and the cup holder.

9. After the adhesive/solvent has set and the parts are securely held together, drill a ⅛-inch-diameter hole in the top of the cup carrier assembly as shown in Figure 5.8.

Figure 5.8: Drilling a hole in the cup carrier

10. Now glue the 3½-inch leather square to the bottom of the Coffee Cup Carrier assembly (see Figure 5.9).

Figure 5.9: Gluing the leather square

Making the Möbius Strip Coffee Cup Carrier Handle

Follow these steps to make the coffee cup carrier handle:

1. Insert the 8-inch soft leather strip through one of the split rings.

2. Turn the leather into a Möbius strip by matching up the ends of the strip and making a half twist so that the top of one end matches up with the bottom of the other end.

3. Overlap the ends by about ¾-inch and fasten them securely either with needle and thread (see Figure 5.10;

use a thimble to push the needle through the leather) or with a couple of pop rivets.

Figure 5.10: Leather Möbius strip

4. Now insert the split ring with the Möbius strip into one end of the barrel swivel.

5. Gently pry apart the other split ring and insert the other end of the barrel swivel.

6. Now insert the split ring into the ⅛-inch hole on the cup holder assembly.

Your Möbius strip coffee cup carrier (see Figure 5.11) is ready to use.

Figure 5.11: Completed cup holder

Using Your Möbius Strip Coffee Cup Carrier

> ✒ **Note** Use clear water with your Möbius strip coffee cup carrier until you get a feel for how it handles.

Now that you've assembled your coffee cup carrier, it's time to put it to good use:

1. Place a paper or plastic cup filled with water on the leather pad of the Möbius strip coffee cup carrier.

2. Place one or two fingers through the leather handle. Begin gently swinging the cup holder in small to medium circles. It doesn't really matter if the circles are large or small—you can even change directions—the water should stay in the cup.

   Keep in mind that the Möbius strip coffee cup carrier will splash or spill its contents if you jerk or whip the handle suddenly, but for most types of smoothly executed motion, the water will remain in the cup (see Figure 5.12).

Figure 5.12: Anti-spill coffee cup holder in action

How Does the Möbius Strip Coffee Cup Carrier Work?

To understand how the coffee cup carrier you made works, you must first consider why liquids spill. The physics behind the coffee stains on your carpet is a bit complex. Not only is an intricate interplay of accelerations, torques, and forces at work, but the biomechanics of bipedal human motion also come into play.

Researchers have found that the wave-like motion of coffee in a mug possesses a unique natural frequency that is related to the size of the mug. Typical coffee mugs produce oscillations that closely match the motion a person makes when walking. Now, if you walk, coffee cup in hand, at a steady and even pace, there is no spillage. But even small irregularities in your gait cause breathtakingly complex accelerations of liquid molecules that even Dr. Möbius would find a challenge to mathematically model. These irregularities amplify the liquid oscillations, which leads to sloshing, and ultimately, to stains on your carpet.

But if you add in the coffee cup carrier with its flexible handle, the math becomes much simpler. Because the handle flexes easily, all the forces acting on the coffee must act in line with the handle. As long as the flexible handle stays taut, there are no lateral (that is, side-to-side) accelerations, and the forces acting on the coffee merely push it toward the bottom of the cup holder. But if the handle bends, say due to a quick change in direction, the whipping action will cause the coffee to spill.

But you may be wondering, what does the Möbius strip have to do with all this? The handle, because it is twisted and tied into a Möbius strip (see Figure 5.13), resists kinking and snarling and makes the carrier even more spill resistant. Another great use for a Möbius strip!

By the way, here's a joke that you'll get, but will your friends? Maybe not!

Why did the chicken cross the Möbius strip? To get to the same side.

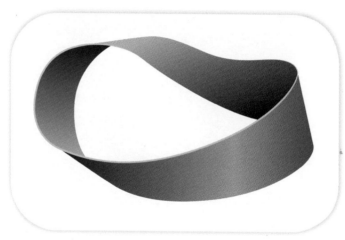

Figure 5.13: The useful Möbius strip

Mathematicians refer to the Möbius strip as a nonorientable surface, meaning it has only one side. That's what makes it so unusual—most surfaces, like a piece of paper (a plane), a globe (a sphere), or a donut (a torus), are orientable. That means, basically, you can paint one side of the surface one color and the other side (or, in the case of the torus or sphere, the inside) another color.

But that's not the case with a nonorientable surface. There is no difference between the inside and the outside or the front and the back.

The Möbius strip is an interesting mathematical curiosity, but it also has practical applications. Imagine a long conveyor belt in a factory. A regular conveyor belt would wear on the inner facing surface where the roller grips the belt, while the outer surface stays relatively pristine. But if the belt is twisted in a Möbius strip, the wear is distributed over both sides and lasts twice as long.

Part II

Inspired by Hephaestus

Hephaestus, the Greek god of the blacksmith's fire and the mythological progenitor of all engineers and tinkerers, was something of a nerd. He was, wrote Homer, extremely intelligent, diligent, and friendly toward mortals, but he was quite unlike the other gods up on Olympus—he did not have much in the way of elegance or style.

Hephaestus was the only non-perfect immortal since he was crippled when Zeus threw him from the mountain after an argument. According to the mythological texts, Hephaestus fell for nine days and nine nights, until he landed on the island of Lemnos. With his leg smashed and mangled, he consoled himself by building a palace next to a volcano. For nine years he worked there, perfecting his engineering and fabricating skills before ascending Olympus again to take his rightful place among the Olympians.

A great craftsman, Hephaestus was the first being, mortal or immortal, to understand how useful technology could be. As we saw in the first book in this series, *ReMaking History Volume 1: Early Makers*, the Hephaestus-inspired ancient Greeks were extraordinarily good at inventing things. But when classical Greece fell, metaphorically speaking, this god fell into a long slumber from which he did not rouse until the 19th century's Golden Age of Invention. Once awake, though, he taught the inventors of the time what he long knew—that technology could turn mortals into gods, at least of a sort.

During this great century of invention, Sir George Cayley invented a machine that would give humans a chance to fly with Nike and Hermes. Joseph McKibben built machines that were, like Hercules, strong enough to move immense weights. And like Hephaestus himself, Hans Goldschmidt found he could command the power of fire.

Hans Goldschmidt

Hans Goldschmidt and the Invention of Thermite

6

One of the most powerful chemical processes used in industry is called the *thermite reaction*. Thermite is a mix of two common chemicals—iron oxide, better known as rust, and powdered aluminum. When these chemicals are combined and ignited, the mixture burns hot enough to melt iron. Wilhelm Ostwald, a Nobel Prize–winning chemist, called the reaction "a blast furnace that fits in a vest pocket" because you can use it to produce overheated liquid steel within a fraction of a second wherever and whenever you have the need.

Goldschmidt Discovers Thermite

Dr. Hans Goldschmidt, a 19th-century German chemist and a student of Robert Bunsen (the man who invented the eponymous Bunsen burner), discovered the powerful thermite reaction in his laboratory. An able chemist, Goldschmidt was given credit for several important contributions to modern chemistry. His most important discovery was almost certainly thermite, which was applied to building railroads in several ways.

Through the 1920s, work gangs built railroads by laying 39-foot-long rails end to end, and then joining the tracks using steel connectors called fishplates and several thick metal bolts (see Figure 6.1). However, mechanically joined tracks made an irritating "clack-clack-clack-clack" sound as the trains rode over them. More importantly, bolted connections become loose, requiring a great deal of ongoing and expensive maintenance. Executives of the day actively searched for a way to reduce the cost of maintaining the thousands of miles of track they owned.

Figure 6.1: Railroad workers installing rails

In 1893, Goldschmidt accidentally developed the process for welding thick sections of steel together in the field. In his laboratory in Berlin, while searching for a method for purifying metal ores, he discovered that a mixture of iron oxide and aluminum would generate 3000°C heat, which was more than hot enough to weld a steel track. He quickly switched his attention to refining this process, which he named *thermite welding*. The process was first used to weld streetcar tracks in Essen, Germany, and within a few years thermite welding was being used nearly everywhere railroad tracks needed to be joined. Needless to say, this discovery made Goldschmidt a wealthy man.

Thermite is still commonly used to repair existing track and, in some cases, build new track. In modern thermite welding, workers butt two pieces of track together, then cover the joint with a heatproof fixture called a *joint mold*. A worker pours thermite powder into a funnel-shaped crucible atop the track. A pan positioned next to the mold catches slag or waste products from the reaction. A combination of chemicals is inserted into the crucible to start the reaction, and then everybody stands back as the thermite does its thing (see Figure 6.2). When the welding is complete, the mold is removed and any excess weld is knocked off with a hammer.

In addition to its uses in welding metals together, thermite has another use. Because of its ability to melt steel quickly, the military frequently uses it to disable captured weapons such as artillery pieces and trucks by melting key metal parts such as cannon barrels and engine blocks.

Figure 6.2: Railroad workers using thermite to repair track

Materials

(2) 2"-diameter steel balls
(The cheapest and easiest
way to get these is to pur-
chase trailer hitch balls.
They generally cost around
$5 to $10.)

Heavy-duty aluminum foil

Gloves

Optional:

3% solution of hydrogen
peroxide (available at all drug
stores)

Hydrochloric acid (sold in
hardware stores often under
the name of muriatic acid)

Tools

Bench vise

Orbital sander, sandpaper, or
angle grinder

Disposable brushes

Rag

Heat gun

Producing the Thermite Reaction

As you might have surmised from the earlier description of a thermite reaction, making and using thermite is not normally a project suitable for amateurs. But in this chapter, we will explore a way to make it in a manner that is safe enough for a junior high science student to undertake with supervision.

The general process is this: 1) form an iron oxide surface layer on two large iron balls; 2) cover one of the balls with a layer of aluminum foil; and 3) clack the two balls together briskly to initiate the chemical reaction between the iron oxide and aluminum.

To run your thermite reaction, follow these steps:

1. Clamp the threaded rod portion of one of your hitch balls into a vise.

2. Using an angle grinder (faster, easier), a hand file (see Figure 6.3), or sandpaper (much slower and harder), remove the rust-resistant plating that covers the iron.

> **Warning:** Angle grinders throw out a lot of hot metal bits as they work, so wear eye protection, leather gloves, and a face shield.

Figure 6.3: Removing the rust-resistant plating with a hand file

The ball surface doesn't have to be particularly smooth when you're done, but all of the plating should be removed.

3. Repeat steps 1 and 2 for the second ball.

4. Rust the surface of the iron balls by immersing them in saltwater for several days (see Figure 6.4); then allow them to air dry in a dark, warm spot.

   (Steps 5–8 are optional.) To speed up the oxidation of iron, you can attempt to jump-start the rusting process using chemical means.

5. Don rubber gloves and eye protection; then paint a thin coat of acid on the iron to clean it (see Figure 6.5).

Figure 6.4: Rusting the balls

Figure 6.5: Painting the ball with hydrochloric acid to hasten the rusting process

6. Let the acid evaporate and then apply the hydrogen peroxide solution to the surface with a brush or rag.

   Almost immediately, the iron will begin to oxidize and a thin coat of rust will appear.

7. Unfortunately, the rust layer is too thin to enable the reaction. Since the goal is to obtain a thick, even layer of rust on the ball, continue to apply the hydrogen peroxide to the iron.

8. You can speed up the oxidation process by applying heat from a hot air gun (see Figure 6.6). If you do this, don't touch the metal until it has cooled.

Figure 6.6: Applying hydrogen peroxide while heating the surface with a hot air gun

Note The rust layer on the ball must be quite thick in order for the thermite reaction to occur. I attempted several methods for speeding up the rusting process (acid, electrolysis, oxygen-rich environments, and so on) with middling success. In the end, the most successful method was simply waiting a week for a reasonable coat of rust to build up.

9. When the balls are thoroughly coated in rust, wrap one iron ball in heavy-duty aluminum foil (see Figure 6.7).

Figure 6.7: One ball wrapped in aluminum foil, the other well coated with rust

10. Don heavy gloves. Grasp the uncovered rusted iron ball by the threaded rod in one hand and hold the aluminum foil-wrapped ball, also by the threaded rod, in your other hand.

11. Carefully, but forcefully, bring the unwrapped, rust covered ball down smartly on the aluminum foiled-wrapped ball.

    Glancing blows produce extraordinary sparks and a loud, firecracker-like snap. If you examine the balls after you strike them together, you will see that the aluminum foil has actually welded to the iron.

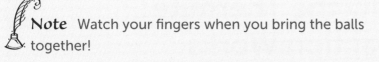

Note Watch your fingers when you bring the balls together!

12. To get a fresh surface of iron oxide, rotate the rusted iron ball after each strike.

   When you strike, aim for the areas with the thickest coating of rust. Producing sparks may seem be a bit tricky at first, but once you get the hang of it, you can put on quite a show, especially if you dim the lights (see Figure 6.8)!

Figure 6.8: Sparks fly when you strike the balls together.

How the Thermite Reaction Works

You can't just lay one ball atop the other and expect to get the spark and crackle of the thermite reaction. Besides the iron oxide and aluminum, you must add energy by banging the balls together in order to start the reaction.

In 1889, Svante Arrhenius of Sweden first used the term *activation energy* to describe the minimum energy that must be added to a collection of chemicals in order to start a chemical reaction. When using a wooden match, for example, the friction of the match sweeping across a rough surface provides the activation energy needed to start the reaction between the phosphorus and sulfur, which causes the match to burst into flames.

The thermite reaction, like a match head, requires some energy to get it started (see Figure 6.9). It's like investing in the stock market: you invest a little of something in order to get a bigger something later.

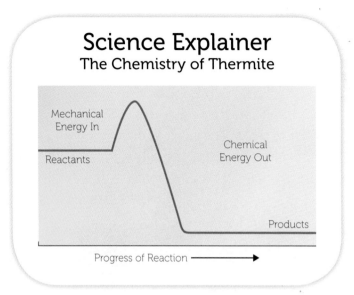

Science Explainer
The Chemistry of Thermite

Mechanical
Energy In

Reactants

Chemical
Energy Out

Products

Progress of Reaction ⟶

One aspect that makes thermite so interesting is the incredible simplicity of the chemical reaction:

$$Fe_2O_3 + 2\ Al - 2\ Fe + Al_2O_3 + \text{lots and lots of heat}$$

Iron oxide (better known as rust) and powdered aluminum are stable at room temperature, and even if you mix purified reagents together, they will just sit there as a mound of inert gray powder. But if you can initiate the chemical reaction with an extremely hot flame (hotter than the flame of a propane torch), the stuff burns wildly in an intense exothermic reaction hot enough to burn through dirt. And once started, there is no way to stop it.

When the thermite reaction occurs, oxygen atoms are ripped out of the iron oxide, which then becomes pure metallic iron. It takes a great deal of energy to accomplish this as the oxygen is bound tightly within the rust molecules. The energy comes from aluminum, a powerful reducing agent that provides enough energy to heat the iron far past its melting point; enough to spew out sprays of sparks and melt adjacent hunks or iron.

This reaction is a bit like riding a bicycle up a small hill in order to get to a much larger and longer downhill run on the other side. The energy required to initiate the thermite reaction (called the activation energy) comes from the mechanical energy of swinging one ball against another. The resulting heat energy, visible as sparks and melted puddles of pure iron, comes from the chemical reaction.

Figure 6.9: The thermite reaction explained

The Brothers Lumière

The Brothers Lumière and the Motion Picture

7

Just as with the telephone, the radio, and the light bulb, controversy swirls around who most deserves credit for inventing motion pictures. Thomas Edison is one possibility, as is Englishman Eadweard Muybridge and Frenchman Étienne-Jules Marey. Each man has his advocates. Mostly, who invented the motion picture depends a bit on how you define the *motion picture*, but a pretty good case can be made for the French brothers Auguste and Louis Lumière being the fathers of the modern movie. Moving picture cameras and display systems existed prior to the Lumières' creations, but the experience we get at the local multiplex is due, in large part, to this pair of brilliant Frenchmen.

The Rise of the Lumières

The Lumières were born in Besançon, France, in the 1860s. Their father, Antoine, was a well-known portrait painter who, seeing an opportunity in the nascent photography industry, started a photographic equipment manufacturing business. Both boys worked for their father: the younger brother, Louis, in a technical role, and the older, Auguste, as a manager. The business became the largest manufacturer of photographic plates in Europe.

In 1894, Antoine saw a demonstration of Edison's kinetoscope, an early motion picture player in which spectators saw moving images by pressing their eye to a peephole (see Figure 7.1). He was impressed—not so much by the technology, but by the potential for an entirely new entertainment-based business. Excited by their father's vision, the brothers took up the challenge of building something that would provide a better, more immersive motion picture experience than the peephole viewing method that Edison's kinetoscope used. If moving pictures were to become popular, then the image, they believed, had to be projected on a large screen. That way, not only could many people watch (and pay for) the movie at one time, but the experience would be bigger, grander, and more exciting.

The brothers moved forward with astounding speed after they got this idea in their heads, designing and patenting their cinématographe in 1895. Unlike Edison's heavy and expensive machine, the Lumières' cinématographe was compact

Figure 7.1: The kinetoscope

and lightweight, weighing a mere 16 pounds. Their design used a simple hand-cranked mechanism (see Figure 7.2) instead of the kinetoscope's heavy, noisy, and expensive DC electric motor. But certainly the most important advance was that their machine was a bona-fide projector, able to throw a moving image onto a large screen, making it possible for a group of people to watch a movie at the same time.

In March 1895, the brothers screened a short film called *La Sortie de l'Usine Lumière à Lyon* for a Parisian audience. The 47-second-long

Figure 7.2: The Lumières' cinématographe

movie (you can see it on the Internet by visiting *www.youtube.com/watch?v= HI63PUXnVMw*) is a bit short on plot and character development. The movie consists of mostly female workers walking out of the gates of the Lumière factory in Lyon, France. The highlight is when a man almost rides a bicycle but then doesn't. *The Godfather* it isn't, but in terms of movie technology, it was a world changer.

The cinématographe, or cinematograph in English, did a lot of things much better than did the praxinoscope, the kinetoscope, the mutoscope, or any of the other motion picture machines that preceded it. The patent drawings detail, among other innovations, its most important technological advance—the incorporation of a sophisticated mechanism called a *cam* to position a single frame of the film stock in front of the projector lens, hold it there for one sixteenth of a second, and then quickly advance the film to the next frame.

If you were to visit an 1895 movie theater, you would doubtlessly notice a projectionist in the back of the room, carefully cranking the cinematograph at a speed of two revolutions per second. By maintaining this cranking speed, he moves the film past the projecting lens at a feed rate of 16 frames per

Materials

(1) 3½"×3½"×⅜" piece of plywood or hardboard (the follower)

(1) 2"×2"×⅜" piece of plywood or hardboard (the cam)

(1) 5½"×8"×¾" piece of plywood or hardboard (the stand)

(1) 5½"×5"×¾" piece of plywood or hardboard (the stand prop)

(2) ⅜" diameter wood dowels, 2½" in length (the pin and claw)

(2) ⅜" diameter wood dowels, 1½" in length (the cam axle and the cam crank)

(2) ⅜" diameter wood dowels, 1½" in length (the pin guides)

Wood glue; paint (your choice of colors)

Tools

Jigsaw or coping saw

Electric drill with ⅜" bit

Sandpaper or file

second, providing a smooth, realistic depiction of motion, perfect for films of factory employees leaving work, passengers getting on and off trains, babies eating crackers, and other popular turn-of-the-19th-century storylines.

More than a century ago, the Lumières brought the motion picture into world focus when they created the modern movie theater experience. They were the pioneers whose cam-controlled movie projector provided the means not only to tell stories, but to permanently preserve and depict nearly all aspects of life.

Making a Lumière Cam and Follower Mechanism Desk Toy

Now that you know the history behind the Lumières cinematograph, we'll make an interesting desk accessory inspired by this important invention.

When you turn the crank of your toy's cam and follower mechanism, which is the heart of the cinématographe, the claw moves with a peculiar motion that you can adjust by making small changes in the profile of the cam with a file or saw blade. You can perform nearly endless experiments with the shape of the cam to make the follower move in a variety of interesting ways.

Follow these steps to construct your cinematograph.

Part 1: Preparing the Cam Follower

Before you begin, don your safety glasses because you're using tools. To assemble the cam follower, follow these steps:

1. Use a jigsaw to cut a 2-inch by 2-inch square opening, centered in the 3½-inch by 3½-inch by ⅜-inch follower piece, as shown in Figure 7.3.

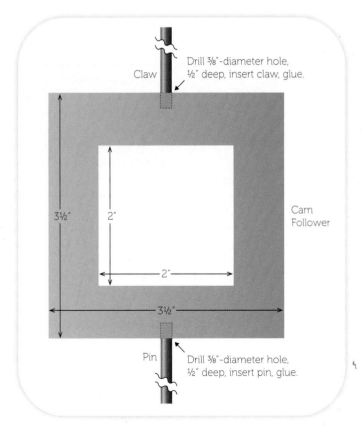

Figure 7.3: Cam follower diagram

2. Use sandpaper to smooth the interior surfaces; they must be completely smooth and free from nicks in order for the cam to slide easily inside the follower.

3. To attach the pin and claw to the cam follower, drill two ⅜-inch-diameter holes about ½ inch deep in the exact center of the upper and lower surfaces of the cam follower (see Figure 7.3).

4. Add a drop of glue to each hole, insert the pins and claw dowels into the holes, and let the glue dry. Figure 7.4 shows what the cam follower should look like at this point.

Figure 7.4: The assembled cam follower

Part 2: Cutting and Fitting the Cam

Now that you've assembled the cam follower, it's time to assemble the cam using the diagram in Figure 7.5 as a guide.

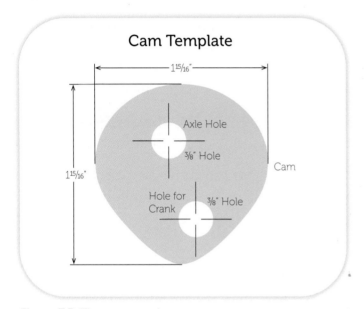

Figure 7.5: The cam template

1. Use a pencil to draw the cam outline depicted in Figure 7.5 on a 2-inch by 2-inch by 3/8-inch piece of plywood or hardboard. Then, cut out the cam using the jig saw or coping saw.

2. After you're done sawing, sand the surface of your cam completely smooth, making sure it is free from nicks or pits.

3. Test the fit of the cam inside the cam follower by placing the cam inside the 2-inch by 2-inch square hole in the cam follower and seeing if you can rotate it completely.

   You must be able to completely rotate the cam inside the hole without any binding or interference. If you find that binding occurs, note where the interference takes place and use sandpaper or a file to remove enough wood from the cam so it turns smoothly.

4. Again take a look at Figure 7.5. Drill the axle hole and the crank hole in the cam as shown in the template.

   The two ⅜-inch-diameter holes should go completely through the cam. Take care to avoid splintering the wood as you drill.

Part 3: Constructing the Stand

Now that the cam and cam follower are complete, it's time to construct the stand that will hold them.

1. Drill three ⅜-inch-diameter holes for the axle hole and pin guide holes ⅜-inch deep in the base, as shown in Figure 7.6.

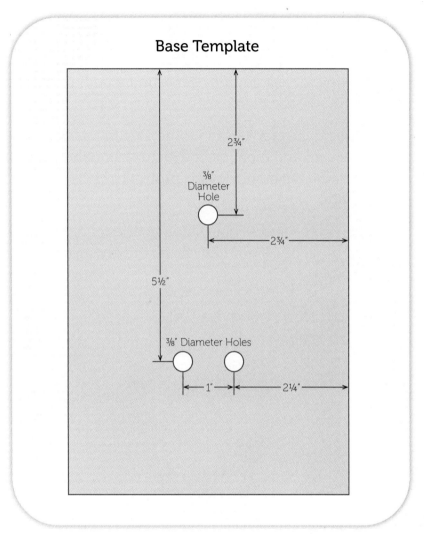

Figure 7.6: The base diagram

Part 4: Attaching the Crank, Base Prop, and Axle and Pin Guides

Now it's time to start putting everything together. Follow these steps:

1. To attach the crank, place some glue in the crank hole in the cam and then insert the crank into the crank hole. Make sure the crank doesn't protrude out the back of the cam.

2. Wipe up any excess glue before it dries and then let everything dry. Figure 7.7 shows the drying crank assembly.

Figure 7.7: Cam with crank added

When the crank assembly is dry, it's time to attach the axle and pin guides to the base.

3. To do so, place some glue in the axle and pin guide holes and insert the axle and pin guides into the holes; let the glue dry.

4. Use glue to attach the prop to the base as shown in Figure 7.8. Again let everything dry.

Figure 7.8: The base with the prop and axle and pins added

5. Before you add the cam and cam follower, carefully smooth all surfaces with sandpaper. The smoother the cam and follower surfaces are, the better they work.

6. Paint if desired.

Part 5: Assembling the Mechanism

Once everything is dry, follow these steps to assemble the mechanism:

1. Place the cam axle through the hole in the cam.

2. Place the cam follower over the cam, making sure the pin is located between the two pin guides as shown in Figure 7.9.

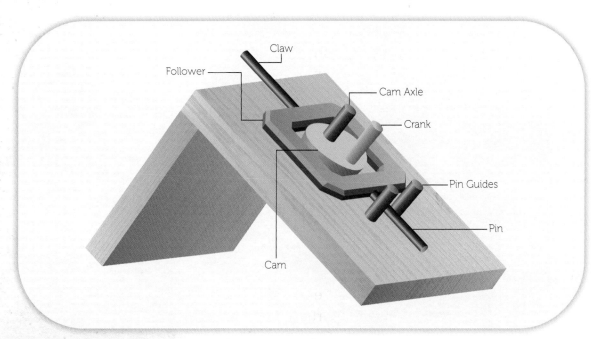

Figure 7.9: Assembling the mechanism

3. Now it's time to turn the crank and try it out!

   As you turn the crank, you will see that the top of the cam follower traces out a repeating motion where the claw rises (in the Lumière projector, this is where the claw engages the perforations in the film stock), moves to the left (where the projector advances the film), remains stationary (*dwells* in engineering lingo) for the $\frac{1}{16}$ of a second where the image is displayed, dips and returns, and then starts over.

4. If the cam and follower seize or bind, reshape the cam to eliminate interference and apply a squirt of powdered graphite to the cam surface to make things turn more smoothly.

5. (Optional) You can also attach a crayon or marker to the tip of the claw to trace out its motion on a piece of paper. If you want to, you can reshape the cam to get different motions that would allow you to use different film sizes or dwell times.

Understanding the Cam

Rotary cams are among the most important mechanisms. In fact, every car and truck has a bunch of them. They are also in sewing machines, weight training machines, machine tools, assembly lines, and myriad other common technologies. Their main purpose is to translate the rotating motion of a spinning object into back-and-forth or linear motion.

As the cam turns, the follower traces out a programmable up-and-down pattern that can be used to control the speeds, accelerations, and directions of machine parts. Figure 7.10 shows such a setup.

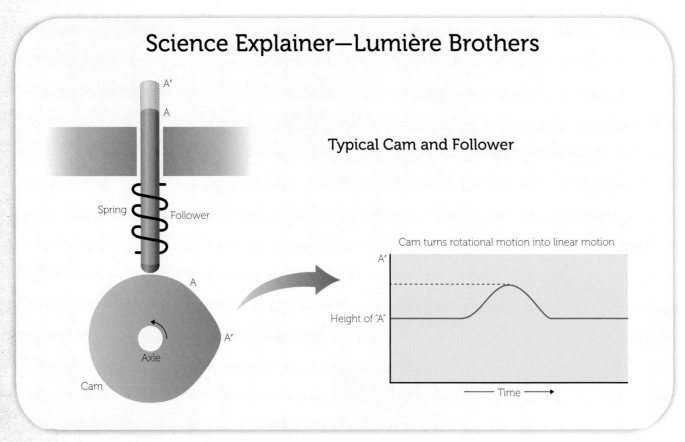

Science Explainer—Lumière Brothers

As the cam rotates around the axle, the spring keeps the follower pressed firmly against the cam surface. The upper surface of the cam follower doesn't move until the follower touches the bump on the cam at position A'. Then it rises vertically until the bump passes. In this way, a rotating cam can execute a repeatable program.

Figure 7.10: Translating rotating motion into linear motion

In their movie projector, the Lumière brothers designed the cam follower to completely enclose the cam. The Lumières' enclosure required neither a spring nor gravity to keep sliding contact between the cam and follower; the mechanism worked flawlessly in any direction or orientation.

The brilliant advance in technology of the cinématographe was the way it used a cam and follower to time the motion of the pins or claws. The cam-controlled pins grabbed the movie film by its perforation, held it stationary in front of the lens for a one sixteenth of a second, and then advanced the film to the next frame.

Chester Rice

Chester Rice and the Loudspeaker

8

In the early days of radio, it was hard to hear the music because the radio waves produced tiny electrical signals. The only practical way to hear these small sounds was to use headphones; basically, you needed to keep the sound-producing diaphragm very close to your ear in order to hear anything. If more than one person wanted to hear the music broadcast, they could place a large horn over the diaphragm. But this acoustic-only horn amplified the sound only minimally.

In an earlier chapter we met Lee de Forest and explored how audio amplifiers work. With this knowledge, engineers started working on ways to boost the sound vibrations coming from a radio and then electromagnetically couple those vibrations to a diaphragm. This device eventually became known as a loudspeaker. Sound systems with loudspeakers could easily fill a whole room with radio music.

Chester Rice at Home

Chester Rice was a gifted engineer and manager for the General Electric Corporation. In the 1920s, Rice lived in Schenectady, New York, in a leafy green part of town known to locals as the GE Realty Plot—a cluster of three-story mansions owned by GE's top executives. Some of these homes had more than six fireplaces, while others had their own ballrooms. Chester Rice's home had something a bit more practical, at least considering his job at GE—it had a fully equipped laboratory.

A large home laboratory suited Rice well. He would start working in his home lab at 11 or 12 o'clock at night and conduct complex experiments until dawn. In the nighttime peace and quiet of the Realty Plot, his ultra-sensitive electronic instruments were undisturbed by activity on the street outside.

Chester Rice's accomplishments were voluminous and covered many areas of electrical engineering. He developed shortwave radio technologies and designed one of the first submarine detection systems. He also advanced the field of radar, testing his embryonic radar gun by measuring the speed of the trolley cars going down the street in front of his house. But arguably, his most important creation was the first really usable loudspeaker, which he developed with fellow GE engineer Edwin Kellogg.

GE's early loudspeaker was unlike anything produced earlier. The unique and wonderful aspect of this speaker was a coil of wire mounted on a stiff diaphragm. When the coil vibrated because of electrical signals going through it, it caused the diaphragm to move as well. The diaphragm's large surface area resulted in much greater sound volumes than had been possible ever before.

Rice and Kellogg's commercially successful loudspeaker was called the General Electric Radiola 104 and it hit the market in 1926. It sold for about $250 at the time, which is well over $3,000 when corrected for inflation. However, the public obviously thought it well worth the money, because thousands of Radiola 104 speakers were sold.

Making a Dynamic Loudspeaker

In this activity, we'll craft a Maker-friendly version of Chester Rice's dynamic loudspeaker. Unlike the Radiola 104, our version will cost $15 or less to make, aside from the cost of 100 feet or so of magnet wire. You will also need to provide an amplifier and a music source.

> **Note** In addition to functioning as an electromagnet, this 28-gauge magnet wire provides resistance to current flowing through the speaker, which is necessary to protect the circuitry of your amplifier. Most amplifiers are designed to encounter 4 or 8 ohms of impedance from the loudspeaker and less than that may damage the amplifier. Twenty-eight-gauge copper wire is rated at about .065 ohms per foot, so 60 feet of it will provide 4 ohms and 120 feet of it will provide 8 ohms. Make sure you use all 120 feet in your speaker. If you use different gauge wire, you will need to change the length accordingly.

Materials

60–120′ of 28-gauge magnet wire

(2) ¾″-diameter × ¼″ cylinder (disc) magnets

(2) Cardboard strips, ½″×1½″

(1) 6″×6″ piece of 1″ (nominal) board

(1) ¾″-diameter dowel, 6″ in length

(1) Diaphragm consisting of a 4″-diameter circle of light, stiff material. Examples include ¹⁄₁₆″-thick balsa wood, a plastic disposable plate, or an empty and clean large-sized tuna fish can. You can experiment to see which material provides the best sound.

Tissue paper

Sandpaper

Quick-setting epoxy glue

Amplifier compatible with 4 ohm or 8 ohm speakers

Music source (digital music player, smart phone, or phonograph)

Making Your Rice and Kellogg–Inspired Dynamic Speaker

Refer to the Figure 8.1 to gain a visual understanding of the overall relationship between the parts of the loudspeaker for this activity. When you're done, it's time to get started!

1. Wrap one turn of tissue paper around the dowel; then wrap the magnet wire around the tissue paper to make a voice coil between ½-inch and ¾-inch wide.

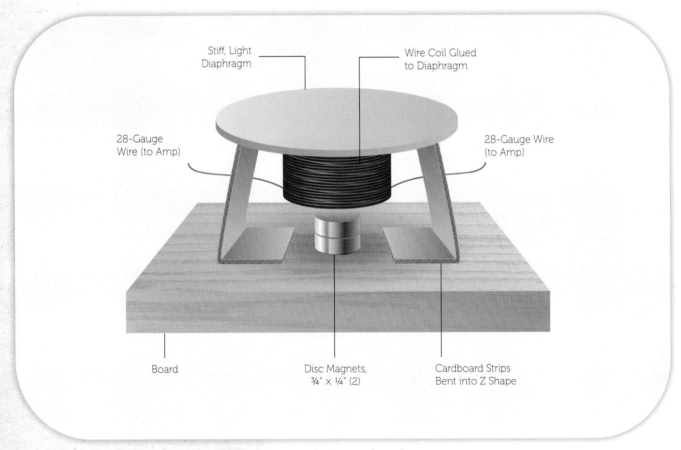

Figure 8.1: The loudspeaker assembly diagram

Leave 12 inches of loose wire at the start and finish of your wrap. Wrap the wire tightly, but not so tightly that you can't slide it off the dowel when you're done wrapping.

2. Gently remove the magnet wire coil from the tissue paper and dowel, taking care that it doesn't unravel.

3. Smear a thin coat of glue on the coil and let it harden (see Figure 8.2.

Figure 8.2: Smearing glue on the coil

4. Once the glue is dry and hard, sand off an inch of insulating paint from the wire ends, as shown in Figure 8.3.

Figure 8.3: Sanding off the insulating paint

5. Now glue this voice coil to the center of the diaphragm as shown in Figure 8.4. Let the glue harden.

Figure 8.4: The voice coil attached to the diaphragm

6. While you're waiting for the glue to dry, fold the cardboard strips into Z shapes as shown in Figure 8.5. Glue the two disc magnets, stacked on top of each other, and the Z-shaped cardboard strips to the 6-inch square board base as shown.

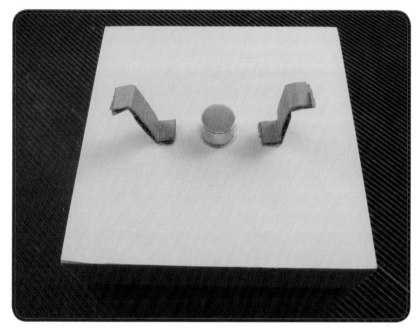

Figure 8.5: Attach the magnet and standoffs to the base.

7. Glue the diaphragm to the other ends of the Z-shaped strips as shown in the Figure 8.6 on the next page.

   Note that the spacing between the voice coil and the top of the magnet affects the quality and volume of the sound. You can experiment to see what height produces the best sound.

Figure 8.6: Adding the diaphragm to the assembly

8. Once you've gotten everything assembled to your liking, connect the speaker wires to your amplifier as shown in Figure 8.7, and enjoy your homemade sound!

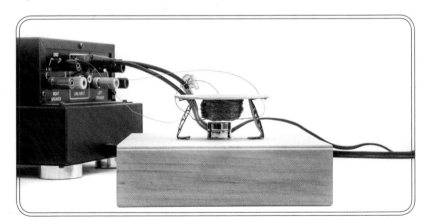

Figure 8.7: The finished speaker

How Does a Loudspeaker Make Sounds?

Sounds travel through the air as waves. When you strike a drum with a stick, for example, the vibrating drum head moves the air molecules surrounding it. These air molecules in turn push on their neighbors, causing them to move also. In this way, a drumbeat moves outward from the drum, creating a sound wave in the air made up of compressions in the spaces between the air particles. But when the wave passes, the air molecules return to how they were spaced, temporarily creating low-pressure regions. Sound waves are simply patterns of high-pressure regions, called *compressions*, and low-pressure regions, called *rarefactions*, traveling through the air.

Like a drum head, the vibrating diaphragm of your speaker also causes waves of high pressure and low pressure to travel through the air (see Figure 8.8 on the next page). When the pressure wave reaches your ear, it pushes your eardrum inward and outward, the middle ear's nerves convert to electrical signals, and your brain interprets this as sound.

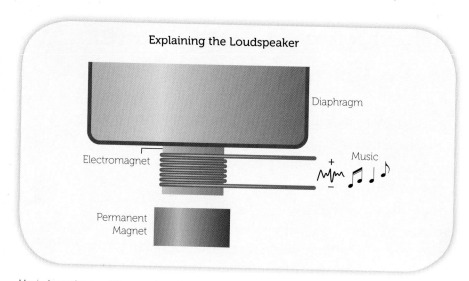

Music from the amplifier consists of electrical signals that vary in amplitude (height) and frequency (width). The electrical signals flow into a thin coil of wire that is attached to a thin, stiff diaphragm, like paper or plastic.

As the electrical signals course through the wire coil, the coil becomes an electromagnet that is pushed away or attracted to the permanent magnet. This causes the diaphragm to vibrate, and the vibrations in air become music.

Figure 8.8: Turning electricity into sound

In your speaker, the center of the diaphragm is glued to a magnetic coil held by the paper strips just in front of a permanent, or field, magnet. When you connect the loudspeaker to your amplifier and music source, small pulses of electricity, which are shaped by the musical notes, travel through the speaker cables into the coil.

Rice and Kellogg's speaker incorporated an electromagnet in which musically shaped pulses of electricity caused the wire coil to produce a pulsing electromagnetic field. As the pulses flowed, the electromagnet either attracted or repelled the permanent magnet, thus moving the diaphragm and creating sound waves.

Joseph McKibben

Joseph McKibben and the Air Muscle

9

As we've seen, the roots of modern engineering emerged in the 19th century. In 1828, an early association of engineers in the United Kingdom, called the UK Institution of Engineers, wrote that the goal of every engineer was to "[direct] the great sources of power in nature for the use and convenience of man."

Since then, engineers and scientists have continued to make wonderful, constructive devices that make life better for people who are sick and disabled. Not nearly as often, they design and build weapons and implements of warfare. And once in a while, the same person does both.

McKibben and the Atom Bomb

Joseph Laws McKibben was an important figure in World War II's Manhattan project. A nuclear physicist from the University of Wisconsin, McKibben was a member of the group of scientists and technicians who researched the atomic properties of the *tamper*, the device that controlled the speed and power of the bomb's chain reaction.

In addition to being a theoretician, McKibben was a hands-on sort of scientist. On July 16, 1945, he was the last man to touch the first ever atomic bomb—code-named Trinity—before it was set off in the New Mexico desert. McKibben's job was to make the final electrical connections to the explosives that would initiate the chain reaction after it was suspended in a harness of wires from a steel tower. After making sure all the connections were secure, he hopped into his jeep and quickly drove two miles to a concrete bunker where the countdown to H-Hour was under-way. There, McKibben pushed the button that initiated the final control sequence that set off the atomic bomb, seen here in Figure 9.1.

Figure 9.1: Explosion of the Trinity atomic bomb

A Human Muscle, Powered by Air

Six years later, in 1952, McKibben's daughter Karan was stricken with polio. Paralyzed from the neck down, she was confined to an iron lung. McKibben felt that he could use his engineering skills to improve the quality of her life. So, working with the doctors at the rehabilitation center where she lived, McKibben began researching ways to give polio patients some control over their fingers.

McKibben began studying the existing hydraulic, electric, and pneumatic methods of moving paralyzed arm muscles, and he became intrigued by one that he felt had particular promise. A few years earlier, a German scientist had prototyped a clever pneumatic gadget operated by compressed gas. The device consisted of a flexible bladder that could be filled with carbon dioxide. The bulging bladder closely simulated the natural motion of human muscles. Could this idea, McKibben wondered, be expanded upon to allow paralyzed fingers to work once again?

Other scientists teamed up with McKibben to develop a prototype. Ultimately, the team developed a workable device that became the gadget now known as the air muscle.

An *air muscle* is an elastic tube encased in a braided fabric covering. Dr. McKibben placed it next to his daughter's paralyzed forearm and attached it to her thumb and first and second fingers with splints (see Figure 9.2). When she operated a lever, which was possible with the limited motion she possessed, gas flowed into the tube, causing a contraction that drew the paralyzed fingers together. At the next touch of the lever, the plastic tube deflated and her fingers relaxed.

KARAN AND PARENTS leave Rancho Los Amigos Hospital where later model of muscle was fitted. Unable to walk, she does high school work at home

Figure 9.2: McKibben's daughter, Karan, with the air muscle

More formally known as a *braided pneumatic actuator*, McKibben's air muscle has become an important pneumatic component, used by roboticists and biomedical engineers. These pneumatic actuators convert pneumatic power to pulling force. They are useful in many applications because they have a high force-to-weight ratio, a flexible structure, and don't cost much to manufacture.

A typical modern air muscle consists of a rubber tube that's surrounded by a braided fabric mesh sleeve. When the rubber is inflated, the mesh expands outwardly around the long axis but simultaneously contracts in length. This shortens the muscle, and anything attached to the ends of the muscle is pulled together. The muscle contracts smoothly and with a surprising amount of force.

Inside the typical air muscle, the expanding volume of air inside a flexible bladder is constrained by the polymer strands of a high-tech fiber sleeve. The pulling force of the air muscle is a function of the air pressure applied, the length and diameter of the muscle, and the material properties of the mesh sleeve and bladder material.

Making an Air Muscle

In this activity, we will construct an air muscle from easily procured hardware store materials. This project is an example of how basic pneumatic principles can be combined with high-tech materials to make devices for controlling motion. It is also a testament to Joseph McKibben's ability to engineer solutions to many different sorts of problems.

> 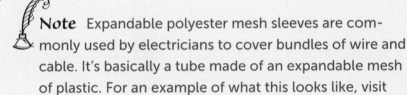 **Note** Expandable polyester mesh sleeves are commonly used by electricians to cover bundles of wire and cable. It's basically a tube made of an expandable mesh of plastic. For an example of what this looks like, visit www.mcmaster.com/#9284K4.

Refer to the right side of Figure 9.3 (on the next page) to get an idea of what you're shooting for and then follow these steps to begin assembling your air muscle:

1. Insert the #10 plastic bolt into one end of the silicone rubber tube. It should fit snugly.

2. Insert the rubber tube inside of the braided sleeve.

   It may take quite a bit of wiggling of the tube and sleeve in order to move the tube inside the slide.

3. Push the middle port of the barbed three-way valve fitting into the open end of the silicone rubber tube as far as possible.

Materials

(1) Soft silicone rubber tube, ⅜" outer diameter (OD), ¼" inner diameter (ID), 8" long

(1) Hard-sided polyethylene tube, ⅜" OD, 2–3' in length

(1) Expandable polyester mesh sleeve, ½" ID, 8" in length (see *www.mcmaster.com/#9284K4.*)

(1) #10 plastic bolt

(1) Box of small (#04) hose clamps

(1) 3-port PVC ball valve with barbed ends for ¼" tube (see *www.mcmaster.com/#4757k57*)

(1) ¼" NPTF male end to ¼" barbed connector (*www.mcmaster.com/#5372K112*)

(1) Industrial hose coupling plug, with ¼" NPTF female end (*www.mcmaster.com/#6534K56*)

Tools

Scissors to cut the tubing and sleeve to the correct length

Screwdriver

Air compressor or other source of high-pressure air

Teflon tape

Safety glasses

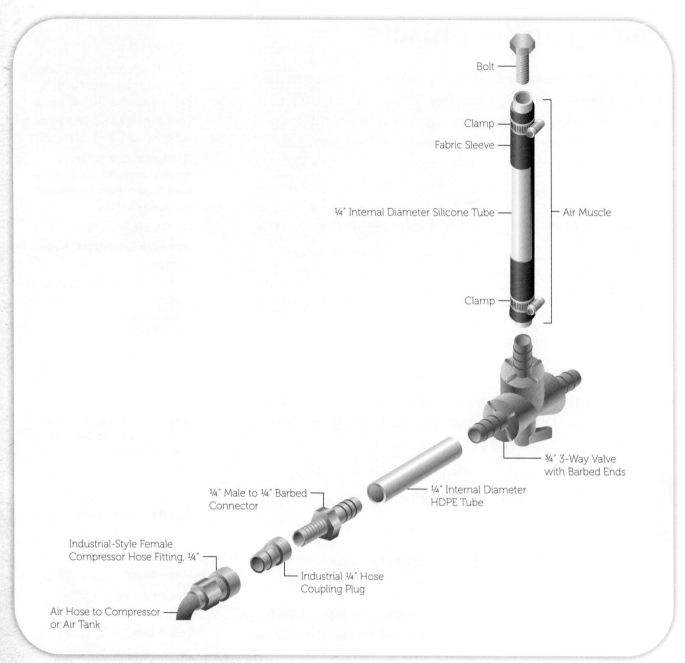

Bolt

Clamp

Fabric Sleeve

¼" Internal Diameter Silicone Tube

Air Muscle

Clamp

¾" 3-Way Valve with Barbed Ends

¼" Male to ¼" Barbed Connector

¼" Internal Diameter HDPE Tube

Industrial-Style Female Compressor Hose Fitting, ¼"

Industrial ¼" Hose Coupling Plug

Air Hose to Compressor or Air Tank

Figure 9.3: The air muscle assembly diagram

4. Place one hose clamp over the sleeve, tube, and barb and tighten securely.

5. Place the other hose clamp over the sleeve, tube, and bolt, and again, tighten securely so it is air tight. This completes the assembly of the air muscle. Now, we'll connect the air to make the muscle move.

6. Cut a 6-inch length of the polyethylene tube.

7. Insert one end into one of the non-center-barbed ports of the three-way valve. Secure the connection with a hose clamp.

8. Insert the barbed end of the ¼-inch barbed male end connector into the other end of the polyethylene tube. Secure with a hose clamp.

9. Connect the female end of the ¼-inch NPTF hose coupling plug to the male end of the ¼-inch NPTF to the barbed connector. Use pipe compound or Teflon tape to secure the male and female connections.

   Your air muscle is ready for use (see Figure 9.4 on the next page).

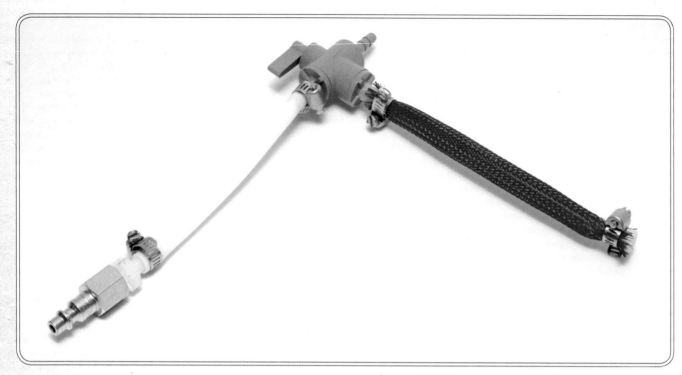

Figure 9.4: The assembled air muscle

Mounting and Using Your Air Muscle

There are many ways to mount your air muscle. The easiest way is to use wire to attach one end of the muscle to a support and the other end to something you want to move. By using levers and pulleys, you can obtain very sophisticated movements for robotics, prosthetics, and automation projects.

To use your air muscle, follow these steps:

1. Connect the air muscle to a high-pressure air source such as a compressor or an air tank. The higher the air pressure, the more the air muscle will contract, but be careful—too much pressure may split the tube.

2. Connect your air source to the air hose coupling plug (see Figure 9.5).

3. Move the three-way valve handle so the air muscle fills with air. As it fills, the silicone tube expands, but it is constrained by the mesh sleeve, which causes the air muscle to contract (see Figure 9.6).

Moving the valve handle in the other direction exhausts the air from the muscle, allowing it to relax.

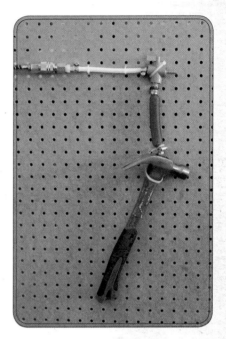

Figure 9.5: Air muscle lifting and lowering a weight (in this case, a hammer)

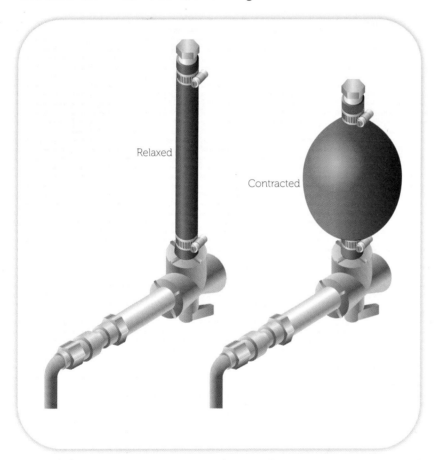

Relaxed

Contracted

Figure 9.6: What happens when the silicone tube contracts

How Air Muscles Work

Why are air muscles so frequently used in robots and artificial limbs? The reason is that no other type of mechanical mover works as close to the way human muscles work as the air muscle. Typically, machines use rotating motors to obtain motion, but the similarities between the muscles and tendons of humans and the elastic and inelastic parts of the air muscle are far closer. In addition, like human muscles, air muscles are compact, and unlike electric motors or gasoline engines, they can start and stop almost instantly.

Figure 9.7 further explains the air muscle.

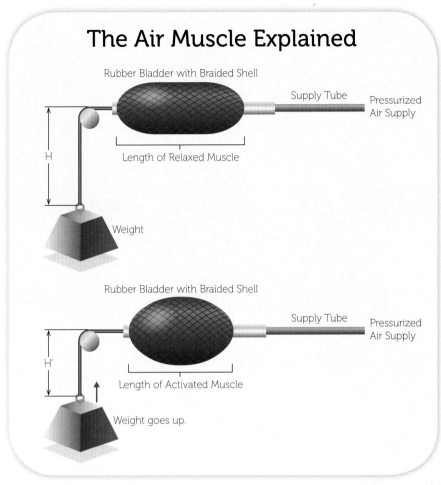

The Air Muscle Explained

Rubber Bladder with Braided Shell

Supply Tube

Pressurized Air Supply

H

Length of Relaxed Muscle

Weight

Rubber Bladder with Braided Shell

Supply Tube

Pressurized Air Supply

H'

Length of Activated Muscle

Weight goes up.

When the rubber bladder in an air muscle is pressurized, it starts to bulge out. But the bladder is encased in a braided shell whose fibers are not elastic. So, the braided shell contracts in length to make up for its newly enlarged volume.

Figure 9.7: How air muscles work

Sir George Cayley

Sir George Cayley and the Glider

10

For centuries, humans have looked with envy at birds for their ability to fly. Even in early times, would-be inventors made wings from feathers or lightweight wood and tested their ability to actually get airborne. But humans are not birds and the muscles of human arms can't compete with the relative strength of a robin's or duck's wing. But that didn't mean people stopped trying. Over time, a lot of great minds took on the challenge. Some were more successful than others.

The title "the father of aeronautics" could be bestowed on a lot of different people depending on your perspective, but I think a very good case can be made on behalf of Englishman George Cayley, the 6th Baronet of Brompton. Sir George was an incredibly gifted inventor who made his mark on many technical fields, including ballistics, civil engineering, biomedical engineering, and mathematics.

Flying without Flapping

Without a doubt, Cayley's greatest achievements were in the field of aeronautical engineering. He did most of his work on flying machines around the turn of the 19th century, about 100 years before the Wright brothers. At the time, there was no engine light enough or powerful enough to get a vehicle airborne, so Cayley never actually achieved powered flight. His most significant inventions were mere unpowered gliders.

Nonetheless, what Cayley did was monumental. His most important contribution? In 1799, Cayley developed the modern concept of the airplane. His idea, like many others that have changed the world, was extremely simple: he was the first to completely separate a flying machine's propulsion systems from its lifting systems. Before Cayley, everyone from Daedalus to da Vinci believed that flapping was the pathway to the air. The flights that had been attempted, unsuccessfully, had involved ornithopters, which were built to use flapping wings that emulated a bird in flight. In Cayley's concept, on the other hand, the lift was provided by a wing that was a fixed, horizontal sail, the surface of which was held taut by the air rushing against it. He called the components that provide thrust the propulsion system, and they were completely separate from the wing and its lifting role.

Another important consideration was Cayley's recognition that many forces operate simultaneously upon a body in flight. The concepts of thrust, lift, weight, and drag, now the cornerstones of aeronautical engineering, were first proffered by Cayley. His ideas prompted him toward real-world tests. In 1804, Cayley built the first recognizable airplane-like thing—a small bamboo and paper glider.

The 1804 glider may not look like much, but believe it or not, it was the first man-made object that incorporated the control and aerodynamic concepts of today's modern airplanes. Cayley's glider had adjustable tail surfaces to control the direction of flight. The observers of the time noted that the glider would "skim for 20 or 30 yards" and "was very pretty to see sail down a steep hill." Bottom line: this thing actually flew!

The Four Forces of Flight

To fully understand Cayley's success, we must consider Cayley's analysis of the forces that act on any flying machine. To make a successful flying machine, four forces must be controlled: lift, weight, thrust, and drag.

- *Lift* is the force that makes the machine rise. Lift in an airplane or glider is caused by the air rushing around the wing.

  Whenever there is a discussion about the technical reasons that explain why airplanes actually rise, the names of Isaac Newton and Daniel Bernoulli always pop up. The actual physics of flight get complicated fast, as several complex fluid dynamic concepts come into play, but for our purposes, it is sufficient to understand that if a wing is shaped correctly, a net upward force on the aircraft body ensues as it moves through the air.

- Counteracting lift is *weight*, which is the force of gravity on the machine.

- To make the machine go forward something must push it, and that force is called *thrust*. Thrust is generated by turning propellers or jets of rearward-directed gas.

- Counteracting thrust is the resistance of the machine through the air, which is called *drag*.

To fly an airplane, a glider, or flying machine of any sort, you must be able to control, contain, and apply these forces.

In 1804, Cayley designed and built the model monoplane glider shown in Figure 10.1.

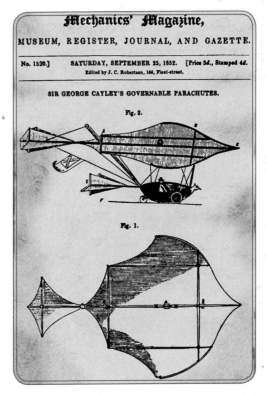

Figure 10.1: The original Cayley monoplane glider

Materials:

2′ of 18-gauge wire

Silk, tissue paper, or Mylar about 2′× 2′

(1) ⅜″×⅜″×36″ piece of spruce or balsa—the fuselage stick

(7) ⅜″×⅛″×3 ¼″ pieces of balsa or spruce—the wing spars

(1) ⅜″×⅛″×6″ piece of balsa or spruce—the center wing spar

(2) ⅜″×1⁄16″×12″ pieces of balsa—tail pieces

(2) ⅜″×1⁄16″×10½″ pieces of balsa—tail pieces

(2) ⅜″×⅛″×36″ pieces of balsa or spruce—wing spanners

(7) cable ties at least 5″ inches in length

(2) ⅜″×⅛″×2″ pieces of balsa—tail block

(1) 2½″ #8 bolt with 3 nuts

Tools:

Drill with 1⁄16-inch drill bit

Thread or ribbon

Hot glue and glue gun

Knife or small saw

More or less modern in appearance, the model featured a fuselage, a kite-shaped wing, and a moveable weight to alter the center of gravity. It was probably the first human-made airplane-like thing that was able to make significant glides.

Cayley flew this model glider successfully in 1804, and if you use your imagination, you can see in it the bones of today's modern airliners. There's a recognizable wing, an adjustable tail and back fins, and a movable weight that could adjust the glider's trim or center of gravity.

Making a Cayley-Inspired Glider

Now that you understand what a glider needs to have in order to fly, let's build a glider very similar to the original Cayley 1804 glider. This glider project has been modified a bit from Cayley's original because, if the truth be told, the original doesn't fly all that well. Making this glider is a matter of cutting small pieces of wood and carefully fitting a strong but light fabric over them to make wings and control surfaces.

I used balsa and spruce to make the fuselage. These are the best woods for model airplane building. Both are strong, lightweight, easily worked, and relatively inexpensive. Traditionally, glider wings are covered in silk. I used a trimmed Mylar plastic sheet because it has better strength-to-weight properties.

The key to making this modified Cayley glider is to insert a slight curve on the upper wing surface to create an airfoil.

An *airfoil* is a curved surface that cuts through a fluid, making a low-pressure section above and a high-pressure section below. This pressure difference provides lift.

Refer to Figure 10.2, which shows the general layout of Cayley's glider.

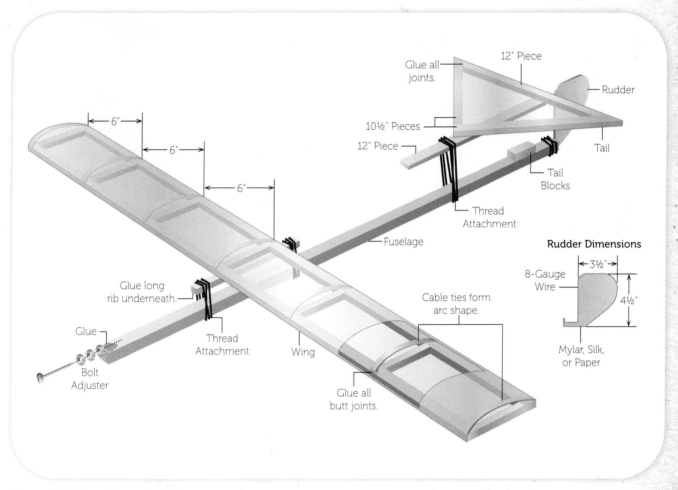

Figure 10.2: The glider assembly

As you can see in Figure 10.3, the wing and tail are attached with thread or ribbons and may be moved about on the fuselage stick to obtain the best flight characteristics.

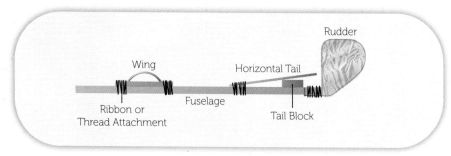

Figure 10.3: Attaching the wing and tail with thread

Part 1: Forming the Rudder

Begin making your glider by forming the rudder:

1. Form the piece of 18-gauge wire into a general rudder-like shape that is about 4½ inches from top to bottom, 3½ inches from side to side, and has a projecting pin across the bottom as shown in Figure 10.4.

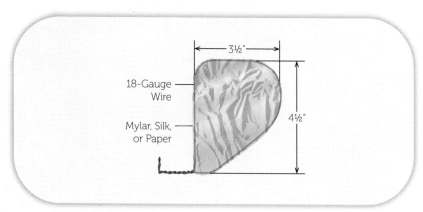

Figure 10.4: Diagram of the rudder

2. When you are satisfied with the shape, clip off any excess wire and then solder the ends of the wire together.

3. Next, you'll need to glue the silk, tissue paper, or Mylar to the wire form as shown in Figure 10.4. To do so, make shallow cuts in the covering at intervals so you can turn the fabric under and glue it neatly. Pull the fabric taut, then glue it into place. When you're done, your rudder should look a bit like Figure 10.5.

4. Now drill a 1/16-inch hole in the main fuselage stick that is just large enough for the rudder's projecting pin about an inch from the end of the fuselage.

Figure 10.5: The completed rudder

Part 2: Making the Tail

Now, it's time to make the glider's tail:

1. First, make sure you have all your balsa/spruce pieces cut and measured as detailed in the materials list.

2. Glue together the triangular horizontal tail piece shown in Figure 10.6 (on the next page) out of the 3/8-inch by 1/16-inch tail pieces.

3. Cover the tail with Mylar or fabric using glue.

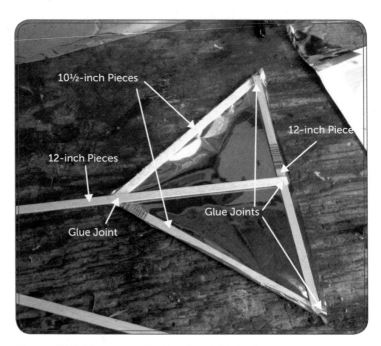

Figure 10.6: The assembled horizontal tail piece

Now it's time to move on to constructing the glider wing.

Part 3: Building the Glider Wing

While the glue is drying on the horizontal tail piece, it's time to assemble the wing superstructure as shown in Figure 10.7:

1. Use the six spars of balsa or spruce measuring ⅜-inch by ⅛-inch by 4 inches, the single ⅜-inch by ⅛-inch by 6-inch center spar, and the two pieces of balsa or spruce measuring ⅜-inch by ⅛-inch by 36 inches to begin assembling the structure. Set this aside and wait for the glue to dry.

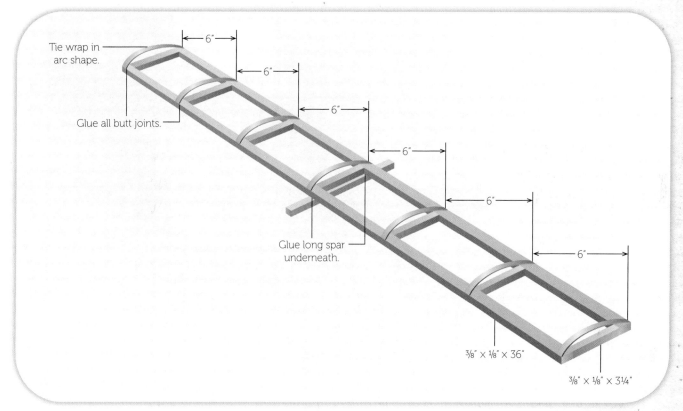

Figure 10.7: A diagram of the wing superstructure

2. After the glue sets, flip the wing structure over and attach the seven cable ties. Hot glue them securely to the 36-inch wing spanner pieces opposite the spars (see Figure 10.8 on the next page). To do so, glue down one end, then form the cable tie into a gentle arc that is about ⅜-inch high at midpoint (you can use a ⅜-inch-wide piece of balsa as a gauge) above the plane of the spars; glue into place.

Figure 10.8: The wing structure with the cable ties attached

3. Once the glue has set, carefully cover the wing with silk, tissue paper, or Mylar using glue. When you're done, the wing surface should be smooth and taut on both sides.

   Refer back to Figure 10.2 for a visual guide to the next three steps.

4. Glue the two ⅜-inch by ⅛-inch by 2-inch-long tail blocks, one atop the other (to make a ¼-inch-high block), to the fuselage underneath the triangular tail.

5. Once the tail blocks are attached to the fuselage, place three nuts on a 2½-inch-long #10 bolt and hot glue the tip of the bolt to the front of the fuselage as shown in Figure 10.2.

Part 4: Assembling the Glider

Now it's time for final assembly of the glider.

1. To attach the rudder, insert it into the tail in the 1/16-inch hole you drilled earlier as shown in Figure 10.3. Wrap the joint with ribbon or thread.

2. Use thread or ribbon to attach the long middle spars of the wings and tail to the fuselage (again refer to Figures 10.2 and 10.3). To begin, place the middle of the wing about 8 inches back from the front of the fuselage. You will likely need to adjust this distance later, during your test flights.

3. Attach the tail to the fuselage as shown in Figure 10.3. The back end of the tail should align with the back of the fuselage stick.

Once you have your wing assembly and tail assembly attached, you are ready to test your glider.

Part 5: Flying the Glider

It's time to fly the glider! All you need to do is hold the glider lightly and give it a level toss.

- If the glider nosedives, untie the ribbon attaching the wing assembly to the fuselage and move the wing back.

- If the glider rises too steeply and then stalls, move the wing forward.

You can make fine adjustments by moving the horizontal tail forward or back and by spinning the nuts on the nose bolt. You can control the yaw (the left or right direction) by adjusting the position of the rudder.

A well-made glider like the one "Sir George" is holding in Figure 10.9 can travel a surprisingly long distance.

Figure 10.9: Sir George is about to launch his glider.

How Do Airplanes Work?

An airplane in flight is a study in balanced forces. If there's not enough lifting force, the airplane glides not so gracefully into the ground. For a plane to attain cruising speed, the thrust from the engines must overcome the friction between the air and the fuselage. Figure 10.10 further describes the forces necessary for flight.

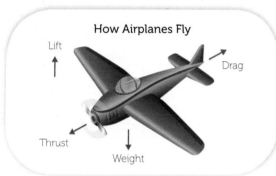

How Airplanes Fly

Lift

Drag

Thrust

Weight

Sir George Cayley was the first to divide the aeronautical forces acting on an airplane into two main parts. The forces that acted horizontally (more or less) to Earth's surface he called *thrust* and drag. Thrust is the force provided by the airplane's engines and *drag* is the resistance the airplane experiences due to air friction.

The force that allows the plane to rise vertically is called *lift*, which is provided by an air pressure differential over the wings as the airplane moves forward. To fly, lift must be greater than the plane's *weight*.

When a pilot can balance all four forces, the airplane will fly smoothly and steadily.

Figure 10.10: The forces of flight

About 50 years after he made his first glider, Cayley returned to glider design and built a big one. In fact, it was big enough to carry his coachman on a 200-yard flight. Why did Cayley order his coachman to be the test pilot? Well, the flight was achieved by towing the glider behind a galloping horse, so who better to be at the reins, so to speak, than his coachman?

The story goes that when the flight was done, the coachman told Cayley, "Sir George, I quit; I was hired to drive and not to fly."

Appendix: Going Further, Deeper, and Higher

After becoming acquainted with the work of the great inventors, engineers, and makers of the Industrial Revolution, perhaps you'd like to read more about their projects. Well, plenty of resources are available! Here are a few ideas:

1. Visit *www.makezine.com*.

   This is the Internet home of *Make:* magazine, which is a bimonthly magazine that provides a ton of do-it-yourself (DIY) ideas in each issue. The projects involve computers, electronics, robotics, metal-working, woodworking, and other disciplines.

2. Visit a Makerspace in your area.

   A Makerspace is an area devoted to making DIY projects. It's a place where people come together to share ideas, tools, and expertise. Visit *http://spaces.makerspace.com/makerspace-directory* to see if there is one near you.

3. Many books are available with additional ideas for exploring science, history, and DIY. My books, *Backyard Ballistics*, *Defending Your Castle*, and *The Art of the Catapult*, are full of interesting projects. Find them and others online or in your local bookstore.

The Inventor's Workshop

If you enjoy making things, eventually you'll need a space to work and tools to work with. In the days of the early inventors, a tool was simply a handheld implement, such as a hammer, saw, or file, and it was used for performing or facilitating mechanical operations, like cutting, pounding, or filing. But in modern times, tools do so much more. They measure quantities and qualities precisely, they join electrical components into circuits, and they perform a hundred other useful operations.

The Workbench

First and foremost, you'll need some sort of flat, solid surface on which to work. Any sturdy table will do, but a workbench is a great help, because it provides the foundation you need in order to work skillfully.

You can make or buy a workbench. Many lumberyards sell prebuilt workbenches or kits containing all the materials you'll need. You can also find a design for one or draw one yourself. Designs for homebuilt workbenches run from complex Scandinavian designs with beechwood frames that are mounted on self-leveling hydraulic cylinders down to a simple plywood door nailed to two sawhorses. No matter what sort of bench you have, the addition of a wood vise and pullout shelf make it more versatile.

Necessary Tools

Ask an expert what sort of tools to buy and the typical advice is to buy the best quality tools you can afford. In most cases, that's good advice. Cheap screwdrivers, for example, can be a big mistake; the soft metal edges of inferior blades can bend or even break under stress, and the plastic handles chip when you drop them. For any tool you use frequently, it makes sense to go with quality.

On the other hand, when you've got a one-off job, and you're not sure if you'll ever have another use for piston ring pliers or a gantry crane, buying an inexpensive tool may make sense.

Here are some ideas for outfitting your workspace.

Basic Tools

These handheld tools are useful in a wide variety of situations. They are as important for adjusting or repairing existing items as they are for making new ones.

Screwdrivers. Choose an assortment of good-quality Phillips-head and flat-head (and, possibly, Torx) screwdrivers in a variety of sizes.

Handsaw. Most often, you'll be cutting dimensional lumber (2×4s, 2×6s, etc.) to size, so choose a saw with crosscut instead of ripping teeth.

Hacksaw. You need this type of saw for those occasions when you have to cut through something harder than wood.

Hammers. Start with a claw hammer for nailing and a rubber mallet for knocking things apart.

Socket and wrench set. If you work on things mechanical, you'll appreciate the quality of a good socket set. Spend the money and get English and metric sockets as well as Allen wrenches.

Pliers. Pliers come in a variety of shapes. At a minimum, your shop should have standard, needle-nose, and vise grips.

Cutters and mat. You'll want diagonal cutters, a utility knife, tin snips, a wire cutter/crimper/stripper, and a good pair of scissors. You'll also find a self-healing cutting mat to be a great help. Buy one at any fabric store.

Clamps. Clamps securely hold workpieces, allowing you to work safely and accurately. Clamps come in various sizes and are selected based on the size of the workpiece.

Linear measuring tools. Make sure you have a tape measure, a protractor, and a combination square.

Files and brushes. You'll need flat and round bastard files and a wire brush. (A bastard file refers to a file that has an intermediate tooth size.)

Mixing and volume measuring equipment. Stock your work area with plastic bowls in different sizes, disposable spoons, measuring cups, and measuring spoons.

Safety equipment. Safety glasses, hearing protection, a fire extinguisher, goggles, a dust mask, and gloves are all very important. All safety glasses, even inexpensive ones, must conform to government regulations, so they all provide adequate protection. However, more expensive ones are more comfortable and look better, making you more inclined to always use them.

Cordless and/or corded electrical drill. A drill with a variety of screwdriver tips and drill bits may well be your most frequently used power tool. Corded drills are lighter and more powerful, but many people appreciate the flexibility of a cordless model. The larger the top-end voltage (e.g., 14.4 or 18 volts) of a cordless drill, the greater its torque and the more it weighs.

Specialty Tools

Inventors often need specialized tools to perform certain tasks. They are typically not expensive, at least for entry-level tools.

Soldering iron. Choose a variable-temperature model with changeable tips.

Magnifying lens. You'll find a swing-arm magnifier with a light to be a very helpful addition to your shop. It mounts directly to your workbench and swings out of the way when not in use. It's great for everything from threading needles to examining surface finishes.

Scale. A triple-beam balance or an electronic scale is a necessity for chemistry projects and mixing stuff.

Digital multimeter. If you do any electronics work, a volt-ohm meter with several types of probes and clips is indispensable.

Power Tools

These are great, if you can justify their cost:

Drill press. A sturdy drill press provides far more accuracy and drilling power than a hand drill.

Belt sander. Belt sanders utilize a rotating abrasive belt to quickly remove material from workpieces.

Grinder. Grinders have rapidly spinning abrasive wheels and are used for shaping metal and sharpening tools.

Table saw/band saw/scroll saw. Electrically powered saws cut wood much faster than handsaws. However, they must be used with great care.

Beyond these basics, there are hundreds, if not thousands, of tools available—all of which may be useful, depending on the project. In regard to stationary power tools, it's a tough call. Because they are expensive and require a lot of shop real estate, it really depends on what you're going to do *most*. I use my table saw all the time, but I know people who consider a band saw to be an absolute necessity, and others who say a scroll saw is their number one power saw priority.

Supplies

Besides raw materials and tools, stock your shop with key general supplies. Here's my suggested checklist:

- Duct tape
- Electrical tape
- Transparent adhesive tape
- Powdered graphite lubricant
- Rope or cord
- String or twine
- Light all-purpose oil
- White glue
- Superglue (cyanoacrylate)
- Quick-set epoxy
- Extended-set epoxy
- Sandpaper: fine, medium, and coarse
- Heat-shrink tubing
- Zip ties
- Pencils
- Ink markers
- Rags, wipes, and towels

It takes time and money to accumulate a good supply of tools. But a well-stocked workshop or tool box and the ability to use the tools properly are valuable assets for any inventor.

Ideas for Further Study

Y ou may not be familiar with the 17th century scholar Isaac Watts, but he played an important role in the advancement of science. Interestingly, he wasn't actually a scientist. He was an English logician, teacher, and musician, and he's probably best known for being the composer of the hymn "Joy to the World."

But aside from all that, he was an important theorist on the nature of learning, an original thinker on the subject of scientific and logical pedagogy. His influence on the inventors and experimenters who appear in these pages was immense.

Watts urged his students to improve their minds, and he wanted them to do more than simply read about a subject. He promoted learning about something by using five different methods, which he called his "five pillars of learning." Watts's five pillars are observation, reading, instruction by lecture, mediation, and finally, hands-on experimentation.

So with Watts's philosophy in mind, here are five ideas for further study.

Ride a Möbius Strip Roller Coaster

Kennywood Amusement Park near Pittsburgh, PA, features a roller coaster named the Racer, which is very likely the largest Möbius strip in the United States. Because the roller coaster designers incorporated a Möbius strip into the track layout, two cars are able to travel side by side for part of the ride, providing riders with the feeling that the roller coaster cars are actually racing one another.

The largest Möbius strip in the entire world is likely another roller coaster called the Grand National, which is located at the Pleasure Beach theme park in Blackpool, England.

Fly a Glider

Experience firsthand what Sir George Cayley's coachman must have felt when he took the world's first glider ride. The activity of sailing through the air in an unpowered glider is called soaring. Licensed soaring pilots flying FAA-certified sailplanes, which are also known as gliders, provide the thrill of staying aloft without an engine. You can soar like a bird and try out the flight patterns of large birds such as eagles and hawks. Many small airports offer the chance to soar. For more information on finding a glider pilot, contact the Soaring Society of America (*www.ssa.org*).

See Samuel Morse's Greatest Artwork

Telegraph inventor Samuel F. B. Morse began his career as a portrait artist in the early 1800s. He tired of portraiture and instead became a painter of historical works. His best-known artwork is called "The Gallery of the Louvre." This is a physically huge work, spanning an impressive 6 feet by 9 feet. The painting depicts 38 Italian and French Renaissance masterpieces at the Louvre in Paris, and Morse hoped this piece of high-concept art would draw paying crowds to view it. At the time, his idea did not make him much money, so he quit painting altogether and focused on developing his idea for the telegraph.

But now Morse's original idea is coming to fruition. "The Gallery of the Louvre" has been rediscovered by the art-going public and is scheduled to be displayed over the next several years in several important U.S. art galleries in a traveling exhibition titled "Samuel F. B. Morse's 'Gallery of the Louvre' and the Art of Invention." You can find out more at *www.galleryofthelouvre.com.*

Walk across a Squire Whipple Bridge

Squire Whipple's truss bridge design was a very important development in civil engineering. One of Whipple's bridges still stands near Vischer Ferry, New York. Built in 1869, this bridge, called a bowstring truss bridge, is preserved in a nature conservancy where it spans a section of the old Erie Canal.

See a Movie at the Institut Lumière

Based in Lyon, France, the Institut Lumière is a museum that promotes and preserves the history of French filmmaking and spotlights the lives and inventions of the Lumière brothers. The museum features a large collection of early filmmaking technology, including the Cinématographe, and of course the museum exhibits movies. The museum is located in a house once owned by the Lumières.

Read a Book

Bill Bryson's book, *At Home: A Short History of Private Life*, describes the impact that technology made on everyday life over the last few hundred years. It's a fast-moving, lively explanation of how the things we use every day—in the kitchen, in the bathroom, and elsewhere in our homes—have changed. Although it doesn't go into great technological detail, it is a entertaining and easy-to-read book.

The Devil in the White City: Murder, Madness, and Magic at the Fair That Changed America, by Erik Larson, is a true-crime book set at the 1893 World's Fair in Chicago. Apart from the gruesome murder story, Larson well describes the incredible technology, architecture, and innovation that made this world's fair one of the best attended events in American history.

Index

Morse code, 35
movie projector, xvii, 107. *See also* Lumière brothers
music source, using for loudspeaker, 111, 118
Muybridge, Eadweard, 93
Mylar, using for glider, 136, 139, 142

N

nails, material for telegraph, 28
Navajo Bridge, 39–40
New York World's Fair of 1939, xvi, 2
nonorientable surface, 75
NPTF male and female ends, using for air muscle, 125, 127
nuts and bolts, using for glider, 136, 142

O

orbital sander, using for thermite reaction, 84
O-rings
    assembly for fire piston, 2
    using for fire piston, 8–9, 11
oxidation process, using for thermite reaction, 86

P

paint, using for cam and follower, 96
Paris Expo of 1889, 2
petroleum jelly, using for fire piston, 8
plastic bolt, using for air muscle, 125
plastic polish, using for fire piston, 8
plate weights, material for truss bridge, 41
pliers, 151
plug, assembly for fire piston, 2
plywood, using for cam and follower, 96
pneumatic actuators, 124
polyester mesh sleeve, using for air muscle, 125
polyethylene tube, using for air muscle, 125, 127
polygon of forces, 38–39
pop rivet gun and rivets, using for cup holder, 66
power tools, 153
Pratt truss bridge, illustration of, 41
propulsion system, 134
prototyping board, using for touch switch, 54, 56
PVC ball valve, using for air muscle, 125
pyrophoricity, 7

Q

Quebec Bridge, 39–40

R

Radiola 104, 110
rag, using for thermite reaction, 84
railroads, construction of, 82–83
rarefactions and compressions, 117
RCA, participation in New York World's Fair, 2
relay, in telegraph diagram, 34
resistors
    for amplifier-based touch switch, 54
    in diagram for touch switch, 55
    in triode, 61
resources, 146–147
resultant, 39
ribbon, using for glider, 136, 143
Rice, Chester. *See also* loudspeaker
    biography, 110
    illustration, 108
    on timeline, xvii
rod
    assembly for fire piston, 2
    material for fire piston, 8
    using for fire piston, 10–11
roller coaster, 158
rotary cams, 105–106
round head screws
    material for telegraph, 22
    in telegraph key diagram, 27
ruler
    material for truss bridge, 41
    using for cup holder, 66

S

safety equipment, 151–152
safety glasses, using for air muscle, 125
safety precautions, xviii
sandpaper
    cam and cam follower, 96, 98, 104
    cup holder, 66
    fire piston, 8, 11

Bring History to Life When You Re-Create It Yourself!

History, technology, and do-it-yourself experience combine in this unique series for Makers. Learn about the great inventions of the past... and then *make* them!

Best-selling author William Gurstelle is a regular contributor to **Make:** magazine. Learn more at *makezine.com*

ReMaking History, Volume 1
Early Makers
ISBN-13: 9781680450606   US $19.99

ReMaking History, Volume 2
Industrial Revolutionaries
ISBN-13: 9781680450668   US $19.99

Master Workshop Tools with Charles Platt's Shop Class in a Book

Charles Platt's *Make: Electronics* has been the best-selling guide to hobby electronics for more than six years. In his latest book, he applies the same entertaining "Learning by Discovery" system to teach the safe, correct, and creative use of hand tools.

Make: Tools
How They Work and How to Use Them

ISBN-13: 9781680452532   US $24.99